# Visions of the Future: **Astronomy and Earth Science**

Leading young scientists, many holding prestigious Royal Society Research Fellowships, describe their research and give their visions of the future. The articles, which have been re-written in a popular and well-illustrated style, are derived from scholarly and authoritative papers published in a special Millennium Issue of the Royal Society's *Philosophical Transactions* (used by Newton; this is the world's longest running scientific journal). The topics, which were carefully selected by the journal's editor, Professor J. M. T. Thompson FRS, include the Big Bang creation of the universe; man's exploration of the Solar System; Earth's deep interior; current scientific ideas about global warming and climate change. The book conveys the excitement and enthusiasm of the young authors for their work in astronomy and earth science. Two companion books cover physics and electronics, and chemistry and life science. All are definitive reviews for anyone with a general interest in the future directions of science.

MICHAEL THOMPSON is currently Editor of the Royal Society's *Philosophical Transactions* (Series A). He graduated from Cambridge with first class honours in Mechanical Sciences in 1958, and obtained his PhD in 1962 and his ScD in 1977. He was a Fulbright researcher in aeronautics at Stanford University, and joined University College London (UCL) in 1964. He has published four books on instabilities, bifurcations, catastrophe theory and chaos and was appointed professor at UCL in 1977. Michael Thompson was elected FRS in 1985 and was awarded the Ewing Medal of the Institution of Civil Engineers. He was a senior SERC fellow and served on the IMA Council. In 1991 he was appointed Director of the Centre for Nonlinear Dynamics.

# Visions of the Future:
# Astronomy and Earth Science

Edited by J. M. T. Thompson

CAMBRIDGE
UNIVERSITY PRESS

PUBLISHED BY THE PRESS SYNDICATE OF THE UNIVERSITY OF CAMBRIDGE
The Pitt Building, Trumpington Street, Cambridge, United Kingdom

CAMBRIDGE UNIVERSITY PRESS
The Edinburgh Building, Cambridge CB2 2RU, UK
40 West 20th Street, New York, NY 10011-4211, USA
10 Stamford Road, Oakleigh, VIC 3166, Australia
Ruiz de Alarcón 13, 28014 Madrid, Spain
Dock House, The Waterfront, Cape Town 8001, South Africa

http://www.cambridge.org

First published 2001

Printed in the United Kingdom at the University Press, Cambridge

*Typeface* Trump Mediaeval 9/13 pt.   *System* QuarkXPress™  [SE]

*A catalogue record for this book is available from the British Library*

*Library of Congress Cataloguing in Publication data*

Visions of the future : astronomy and Earth science / edited by J. M. T. Thompson
    p. cm.
  ISBN 0 521 80537 6 (pb)
    1. Astronomy. 2. Earth sciences. I. Thompson, J. M. T.

  QB51.V56 2001
  520–dc21    00-063032

ISBN 0 521 80537 6 paperback

# Contents

v

**Global warming and climate change**

# Preface

Writing here in a popular and well-illustrated style, leading young scientists describe their research and give their visions of future developments. The book conveys the excitement and enthusiasm of the young authors. It offers definitive reviews for people with a general interest in the future directions of science, ranging from researchers to scientifically minded school children.

All the contributions are popular presentations based on scholarly and authoritative papers that the authors had published in three special Millennium Issues of the Royal Society's *Philosophical Transactions*. This has the prestige of being the world's longest running scientific journal. Founded in 1665, it has been publishing cutting edge science for one-third of a millennium. It was used by Isaac Newton to launch his scientific career in 1672, with his first paper 'New Theory about Light and Colours'. Under Newton's Presidency, from 1703 to his death in 1727, the reputation of the Royal Society was firmly established among the scholars of Europe, and today it is the UK's academy of science. Many of the authors are supported financially by the Society under its prestigious Research Fellowships scheme.

Series A of the *Philosophical Transactions* is devoted to the whole of physical science, and as its Editor I made a careful selection of material to cover subjects that are growing rapidly, and likely to be of long-term interest and significance. Each contribution describes some recent cutting edge research, as well as putting it in its wider context, and looking forward to future developments. The collection gives a unique snapshot of the state of physical science at the turn of the millennium, while CVs and photographs of the authors give a personal perspective.

The three Millennium Issues of the journal have been distilled into three corresponding books by Cambridge University Press. These cover

*Astronomy and Earth Science* (covering the topics described below), *Physics and Electronics* (covering quantum and gravitational physics, electronics, advanced computing and telecommunications), and *Chemistry and Life Science* (covering reaction dynamics, new processes and materials, physical techniques in biology and the modelling of the human heart).

Topics in the present book on astronomy and earth science include the creation and history of the universe according to the Big Bang theory; man's exploration of the Solar System; Earth's deep interior; and current scientific ideas about global warming and climate change.

J. M. T. Thompson, FRS
Editor, *Philosophical Transactions of the Royal Society*,
Centre for Nonlinear Dynamics, University College London

# 1
# Big Bang riddles and their revelations

## João Magueijo[1] and Kim Baskerville[2]

1 *Theoretical Physics, The Blackett Laboratory, Imperial College of Science, Technology and Medicine, Prince Consort Road, London SW7 2BZ, UK*
2 *London Oratory School, Seagrave Road, London, SW6 1RX, UK*

## 1.1 The Big Bang riddles

The Big Bang universe is a success story. It makes use of the general theory of relativity to set up the most minimalistic model for our universe. According to this model the embryo universe was concentrated in a single point, which exploded in a Big Bang event some 15 billion years ago. The Big Bang universe is homogeneous in space, and expands as time progresses: a dynamical prediction of relativity. An elegant explanation for an ever-growing array of observations ensues.

A closer examination of this model, however, reveals a number of unnatural features. The Big Bang universe is fragmented into many small regions, which are so far apart that light, or indeed any interaction, has not had time to travel between them. These 'horizons' are therefore unaware of each other, yet mysteriously share the same properties, such as age and temperature. It almost looks as if telepathic communication has taken place between disconnected regions. Another puzzle is the observed near flatness of the universe. Flatness is central to successful Big Bang models, but is unfortunately not stable. Big Bang models may be open (hyperbolic), flat, or closed (spherical). Closed Big Bang models expand to a maximum size, and then recollapse, dying in a Big Crunch. Open models expand too fast, leaving the universe empty soon after the Big Bang. The problem is that even slight deviations from flatness grow very quickly, leading inevitably to either a catastrophic Big Crunch or emptiness. The fact that neither has occurred means that we are successfully walking on a tight-rope. Short of invoking divine intervention, how can we possibly have managed this for so long?

Thankfully a number of natural explanations have been put forward. In all of these, the riddles plaguing the Big Bang act as windows into new physics. Inflationary models of the universe, which have become a leading idea in modern cosmology, were undoubtedly born out of these puzzles. Inflation is perhaps the simplest addition to the Big Bang which leaves behind a universe without mystery. Another explanation is the so-called pre-Big Bang model. This is inspired by string theory, and explores the possibility of the universe existing before the Big Bang. In the progenitor universe lies the secret of the riddles. The most radical explanation is a recent proposal, involving a revision of the special theory of relativity. According to this proposal light might have travelled much faster in the early universe. The varying speed of light cosmology explains the puzzles solved by inflation and pre-Big Bang models, and maybe some additional riddles, too.

In what follows we review the Big Bang model (Section 1.2), and its riddles (Section 1.3). We then describe solutions to these riddles: inflation (Section 1.4), pre-Big Bang (Section 1.5), and varying speed of light (Section 1.6). We conclude with an assessment of the state of the art.

## 1.2 The bright side of Big Bang cosmology

Cosmology, the study of the universe, was for a long time the subject of religion. That it has become a branch of physics is a surprising achievement. Why should a system as apparently complex as the universe ever be amenable to scientific scrutiny? At the start of this century, however, it became obvious that in a way the universe is far simpler than, say, an ecosystem or an animal. In many ways even a suspension bridge is far harder to describe than the dynamics of the universe.

The big leap occurred as a result of the discovery of the theory of relativity, in conjunction with improvements in astronomical observations. If we look at the sky we see an overwhelming plethora of detail: planets, stars, the Milky Way, the nearby galaxies. At first the task of predicting the behaviour of the universe as a whole looks akin to predicting the world's weather, or the currents in the oceans.

If we look harder we start to see that such structures are mere details. With better telescopes we can zoom out, to find that galaxies, clusters of galaxies, even the largest structures we can see, become 'molecules' of a rather boring soup. A very homogeneous soup, called the cosmological fluid. The subject of cosmology is the dynamical behaviour of this fluid when left to evolve according to its own gravitation. The crucial feature is the fact that this fluid appears to be expanding: its 'molecules' are moving away from each other.

What set the universe in motion? Can physics explain this phenomenon? That was one of the many historical roles played by the theory of relativity. The result is encoded in what came to be known as the Big Bang model of the universe. Here we give a taste of this bewildering theory.

Let us start by assuming that the cosmological fluid is homogeneous and also that at every point all directions are equivalent, that is, we have isotropy (note that homogeneity does not imply isotropy). Isotropy requires that the only possible motion relative to any given point $O$ be radial motion. Imagine a sphere around $O$, and consider the velocity vectors of points on this sphere. Now try to comb this sphere (that is, to add a tangential component to these velocities). There will always be a bald patch, no matter how careful one is. Such a bald patch provides a preferred direction, contradicting isotropy. Therefore, isotropy is a hair-raising experience, allowing only radial motion. Any observer can, at most, see outwards or inwards motion around itself. We shall assume for the rest of the argument that this motion is outwards. The speed at which this motion takes place may depend on the distance and on time, but the function must be the same for all central points $O$ considered. This imposes constraints on the form of this function. Consider three collinear points A, B and C, with B at distance $R$ from A and B (see Figure 1.1). In the rest frame of A, let the velocities of B and C be $\mathbf{v}_1$ and $\mathbf{v}_2$ (top case in Figure 1.1). If we now consider the situation from the point of view of B, C is at a distance $R$ and has velocity $\mathbf{v}_2 - \mathbf{v}_1$ (bottom case in Figure 1.1). But, given homogeneity, what B now sees at C should be what A sees at B. Hence $\mathbf{v}_1 = \mathbf{v}_2 - \mathbf{v}_1$, that is $\mathbf{v}_2 = 2\mathbf{v}_1$. Going back to A's perspective, we now find that points at twice the distance move at twice the speed. More generally $\mathbf{v} = H\mathbf{d}$: the recession speed away from any point $O$ is proportional to the distance. $H$ is called the Hubble 'constant'. It is not really a constant, but may depend only on time.

This is a weird law. Any point $O$ sees the stuff of the universe receding, going faster the further away it is. Let us first simplify life, and ignore gravity, so that these speeds do not change in time. Then a cataclysm must

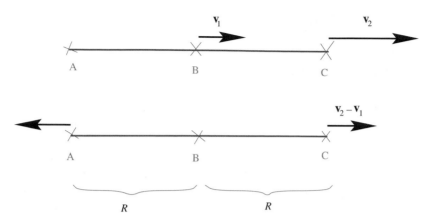

**Figure 1.1.** The recession velocity must be proportional to the distance. First consider the perspective of point A, observing point B recede at speed $\mathbf{v}_1$, and C, located at twice the distance, recede at speed $\mathbf{v}_2$. Then, a comparison with the perspective of point B, and using the fact that B must see A and C recede at the same speed, leads to the conclusion that $\mathbf{v}_2$ must be twice $\mathbf{v}_1$.

have happened in the past. If an object at distance $d$ is moving at speed $v = Hd$, then rewinding the film by $\delta t = d/v = 1/H$ will show that this object was ejected from $\mathbb{O}$. But the rewind time is the same for objects at any distance $d$, and is always $\delta t = 1/H$. Points further away are moving faster, and therefore crossed their greater distance from $\mathbb{O}$ in the same time. Hence, at a time $\delta t = 1/H$ into the past, the whole observable universe was ejected from point $\mathbb{O}$. But $\mathbb{O}$ can be any point. Therefore the whole universe started from a single point, in a big explosion: the Big Bang.

Gravity complicates, but does not essentially alter this argument. And this argument already allows the reader to make pretentious statements about creation. It also allows the reader to share the shock which must have hit Einstein when he first dared apply his theory to the universe as a whole. He was expecting to find a static and eternal universe, in agreement with what was then the observational evidence. Instead he found a restlessly expanding universe, with an explosive Big Bang start. Einstein's equations revealed that the universe had to be in motion!

It is perhaps worth noting that, according to relativity, the universe is not really in motion; rather space itself is moving. In relativity galaxies are not moving away from each other: they are fixed in space, and space is expanding. It is a bit as if the surface of the Earth were expanding at very

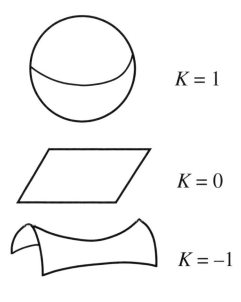

$$K = 1$$

$$K = 0$$

$$K = -1$$

**Figure 1.2.** Big Bang models come in three brands, with distinct spatial geometries. We may have spherical space, with positive curvature; flat or Euclidean space, with which we have been familiar since primary school, and something really weird called a pseudo-sphere, which looks like a saddle.

high speed. The distances between cities would then be increasing in time, but the cities themselves would not be in motion. How much space has expanded is controlled by a quantity called the expansion factor $a$. The expansion factor is a ruler measuring this space that is continuously being stretched. At the Big Bang the expansion factor approaches zero. Nowadays the expansion factor has increased by a factor of $10^{32}$ since the universe became classical, at what we call the Planck time, when the universe was $10^{42}$ seconds old.

Relativity does not uniquely fix the surfaces upon which galaxies are encrusted, and which represent the geometry of our universe. There are in fact three different types of Big Bang model. We label them by a constant $K$, their curvature, which can take values 0, 1, or $-1$. This constant describes the type of curvature of the expanding space. The expanding space can only be a three-dimensional sphere ($K=1$), Euclidean or flat three-dimensional space ($K=0$), or a three-dimensional hyperboloid or saddle ($K=-1$). The two-dimensional analogues are pictured in Figure 1.2.

The brightest side of the Big Bang model is the prediction of universal expansion. What set the universe in motion? The question does not make sense. It's like asking what keeps a free particle moving, as Aristotelian physicists would do. The cosmological expansion is a generic feature of any cosmological space-time satisfying Einstein's equations. Only a restless universe is consistent with relativity; and that is just what was discovered by observation.

## 1.3  The spooky side of Big Bang cosmology

The Big Bang model is a success. It offers the most minimalistic explanation for all the observations currently available. It explains the cosmic microwave background. It explains the abundances of the lighter elements, through a process called primordial nucleosynthesis. It provides an explanation for how structures, such as galaxies, form in a universe which is very homogeneous at early times, indeed at any time at very large scales. This is only to mention a few striking successes of the Big Bang model. Competitors to the Big Bang model, such as the steady-state model, lost their elegance and predictive power as more and more data accumulated.

In the late 1970s, however, it became apparent that not all was a bed of roses with the Big Bang model. Even though the model proved unbeatable when confronting observation, it required a large amount of coincidence and fine tuning, which one would rather do without. These difficulties are referred to as the horizon, flatness, and Lambda problems, which we now describe.

### 1.3.1  Horizons in the universe

Creation entails limitation. Universes marked by a creation event, such as the Big Bang universe, suffer from a disquieting phenomenon known as the horizon effect: observers can only see a finite portion of the universe. The horizon effect can be qualitatively understood from the fact that, since light takes time to travel, distant objects are always seen as they were in the past. Given that creation imposes a boundary in the past, this means that for any observer there must also be a boundary in space. A distance must exist beyond which nothing can be seen, as one would be seeing objects before the creation. Such a boundary is called the horizon.

The existence of horizons is not by itself a problem. The problem is that the horizon is tiny at early times. If we ignore expansion effects, the current horizon radius is 15 billion light years, corresponding to our age of 15 billion years. When the universe was 300 000 years old the horizon radius was only 300 000 light years. If we look far enough we can see the 300 000-year-old universe. As Figure 1.3 shows, we should be able to see many regions which were outside each other's horizons at that time.

The celebrated cosmic microwave background radiation is, in fact, a

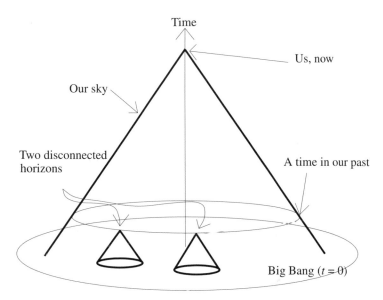

**Figure 1.3.** Choosing to measure time in years and space in light years, we obtain a diagram in which light travels at 45°. This diagram reveals that the sky is really a cone in four-dimensional space-time. When we look far away we look into the past. There is a horizon because we can only look as far as the universe is old. The fact that the horizon is very small in the very early universe means that we can see regions in our sky outside each others' horizon.

glow emitted by the universe when it was 300 000 years old. One can show that a horizon region at this time subtends in the sky an angle of about a degree, twice the size of the Moon. We can see many of these regions, and they all have the same temperature to a high degree of accuracy.

The horizon problem is our ability to see disconnected horizons in our past, and the fact that these horizons are seen to share the same properties, such as temperature or density. The horizon effect prevents any causal equilibration mechanism from explaining this remarkable coincidence. In a sense the horizon problem is really a homogeneity problem: the uncanny homogeneity of the universe across causally disconnected regions.

Expansion complicates this reasoning a bit, but not enough to solve the problem in a Big Bang universe. When we discuss the inflationary universe we shall include the effects of expansion into the discussion, and show how they may be used to solve this riddle, but not with standard Big Bang expansion.

## 1.3.2 Walking on a tightrope without falling off

We saw that there are three different types of Big Bang model, labelled by $K=0$ (flat), $K=1$ (spherical), and $K=-1$ (saddle). Einstein's equations tell us that the actual curvature radius of the universe is $K/a^2$, where $a$ is the expansion factor: expansion stretches the curvature. How does this compare with the way expansion dilutes matter and radiation? Again Einstein's equations provide an answer which is not difficult to understand. For matter exerting no pressure (such as galaxies, which just sit there, and therefore exert no pressure) expansion only pulls matter apart. The dilution in the matter density is therefore just related to the volume expansion, which is proportional to $1/a^3$. Hence the energy density in matter is diluted as $\rho \propto 1/a^3$, the dilution rate corresponding to volume expansion.

Radiation such as the cosmic microwave background, on the other hand, exerts quite a large pressure. Hence it does work as the universe expands. We find that radiation is diluted as $\rho \propto 1/a^4$, the extra factor corresponding to the fact that the radiation pressure is doing work as the universe expands, and therefore depleting its energy more than the naive volume factor. The fact that the energy density in radiation decays faster than that of matter means that the early universe must have been dominated by radiation.

But then we find a distressing fact. Curvature decays even more slowly than either matter or radiation. Hence, if at any time there exist any traces of curvature, these soon come to dominate the universe. The ratio between the contribution to expansion due to curvature, and that due to matter (radiation), increases with $a$ ($a^2$). This means that the flat model is unstable! This conflicts with the fact that we are now close to flatness. Observations show that the contributions to expansion due to matter and curvature are, at most, of the same order of magnitude. How have we managed not to fall off the tightrope of flatness? This is the flatness problem.

To put numbers into the problem, we know that the Big Bang universe has been expanding since the so-called Planck time, $t_p = 10^{-42}$ s when gravity became classical. We can work out that since then the expansion factor has increased by $10^{32}$, or thereabouts, leading to a growth in any deviation from flatness by around $10^{64}$ since then. This means fine tuning the curvature contribution at Planck time by 64 orders of magnitude.

Falling off the flatness ridge is disastrous. Einstein's equations reveal that closed models ($K = 1$) expand more slowly than flat models, but open models ($K = -1$) expand faster. We can follow the evolution of non-flat universes into their curvature-dominated epochs. Expansion in closed models keeps slowing down relative to flat models, until eventually expansion comes to a halt. Once this happens gravity pulls the cosmic matter back on itself, and therefore recollapse starts, retracing expansion's steps, until the universe ends in a Big Crunch. Open models expand ever more rapidly compared to flat models. Therefore, the curvature term keeps increasing until the matter is irrelevant. But this means that the universe is destined to become totally empty. If the contribution to expansion from curvature is not negligible at Planck time, one or the other of these two tragedies would have occurred within a few Planck times.

Only flat models offer a reasonable model for the universe as we see it. But as we saw, they are unstable, the tiniest trace of curvature sufficing to derail them.

### 1.3.3 Einstein's greatest blunder

Self-flagellation has played an important role in modern science. At the start of the twentieth century there was no evidence for cosmological expansion. Relativity predicts expansion, with one exception: a closed universe dominated by a cosmological constant. The cosmological constant represents the energy of the vacuum, and it was introduced by Einstein to ward off expansion. When Hubble discovered expansion, a few years later, Einstein bitterly regretted having introduced the cosmological constant, thus missing yet another theoretical prediction for a major experimental discovery. He called the cosmological constant 'the biggest blunder of my life'.

The cosmological constant may be seen as an extra term one adds to curvature in Einstein's equations. This term is usually represented by $\Lambda$. It may be reinterpreted as an extra fluid pervading the whole universe. It is in this sense that $\Lambda$ is sometimes called the energy of the vacuum. The cosmological constant is the stuff the vacuum is made of. This fluid, however, has a very negative pressure, that is, it is very tense stuff. Recalling the way radiation decays with expansion, how should the energy in $\Lambda$ therefore decay with expansion?

We find the surprising result that the vacuum does not become diluted by expansion! This is because expansion is doing work against the $\Lambda$

tension. Therefore, at the same time as expansion dilutes the energy density in $\Lambda$, it transfers energy into it, via this work. Overall the energy density in $\Lambda$ remains constant.

But this implies that any traces of the cosmological constant would immediately dominate the universe. The ratio between the energy density in normal matter and in $\Lambda$ grows like $a^3$ for matter, and like $a^4$ for radiation. This means a growth by 128 orders of magnitude since Planck time, when we know the universe expansion must have started. We have another, even thinner, tightrope to walk on.

## 1.4 God on amphetamine

In the end, history flushed Einstein's greatest blunder into one of the leading theories of modern cosmology: inflation. Inflation is a period in the early universe during which the dominant energy contribution is the vacuum energy. Inflation is a brief affair with the cosmological constant.

Inflation is a way of switching on the cosmological constant, and then letting it decay into ordinary matter. The trick is played by a field, called the inflaton field. When the inflaton is switched on it dominates all other forms of matter, in the catastrophe described above. However, this catastrophe also brings luck.

One of the predictions of the theory of relativity is that everything, not just matter, generates gravity. Light, heat, energy, everything 'gravitates'. Among these, relativity shows that pressure generates gravity as well. In particular tension, that is negative pressure, generates anti-gravity or repulsive gravity. Hence we may arrive at the conclusion that the cosmological constant, being such a tense 'material', is gravitationally repulsive.

When the cosmological constant dominates the universe we therefore have very fast expansion. Gravity accelerates expansion, rather than acting as a brake. The universe expands exponentially during inflation. A period of inflationary expansion is sometimes also called 'superluminal expansion'. We say that the universe inflates, and this, as we shall see, is enough to solve the flatness and horizon problems.

### 1.4.1 Opening up horizons

In our discussion of the horizon problem we neglected expansion. Let us now refine the argument. The horizon size is the distance travelled by light since the Big Bang. Is this really one light year in a one year old universe?

Travelling in an expanding universe entails a surprise: the distance from the departure point is larger than the distance travelled. This is because expansion keeps stretching the distance already travelled. Imagine a cosmic motorway, realised if the Earth were expanding very fast. Then a trip from London to Durham might show on the odometer that 300 miles were travelled, whereas the actual distance between the two places at the end of the trip would be 900 miles. Similarly in a 15 billion year old universe, light would have travelled 15 billion light years since the Big Bang. However, the distance to its starting point would be roughly 45 billion light years, the current size of the horizon.

This subtlety does not change the essence of the discussion of the horizon effect in Big Bang models, but inflation builds upon this subtlety. With superluminal expansion the distance travelled by light since the start of inflation becomes essentially infinite. Under amphetaminic expansion light travels a finite distance, but expansion works 'faster than light', infinitely stretching the distance from departure.

Therefore inflation opens up the horizons. The whole universe observable nowadays was, before inflation, a tiny bit of the universe well in causal contact. This was then blown up by a period of inflation. We have solved the horizon problem.

## 1.4.2 The valley of flatness

If we insert a cosmological constant into the flatness problem argument, we find a pleasant reversal of the situation. Now the contribution to expansion due to matter (which is $\Lambda$ during inflation) remains constant, whereas the contribution due to curvature decays like $1/a^2$. The ratio between curvature and matter contributions now decreases like $1/a^2$ instead of increasing like $a^2$. Flatness becomes a valley, rather than a ridge.

Because the expansion factor is increasing exponentially, within a very short time any deviation from flatness becomes infinitesimally small. At the end of inflation the contribution to expansion from curvature is smaller than $10^{-64}$. We have achieved the fine tuning required to survive the Big Bang flatness tightrope. Inflation provides the primordial balancing pole to allow us to walk the tightrope without falling off.

## 1.4.3 The end of the nothing

At the end of inflation the inflaton field decays into radiation, in a process known as reheating. The normal course of the Big Bang resumes, but the worst Big Bang nightmares have been staved off. It is no longer a coincidence that the universe is homogeneous across so many disconnected horizons. All these separate horizons went to the same nursery school. The instabilities of the sensible brand of Big Bang models (the flat ones) are no longer a concern. A period of inflation finely tuned the universe. It gave it the stability at birth required for the universe to cope with its 'instabilities' in later life.

The only problem inflation does not solve is of course the $\Lambda$ problem. Inflation is built upon it. If in addition to the inflaton effective $\Lambda$, which turns on and off, there is a genuine cosmological constant, this will still be present at the end of inflation. The energy densities for the two $\Lambda$s remain constant, and therefore at fixed ratio, during inflation. Hence a genuine $\Lambda$ would still threaten to dominate the universe at any time after inflation. Inflation does not provide the fine tuning required to solve the $\Lambda$ instability of the Big Bang.

## 1.5  Is there life before the Big Bang?

There have been several attempts to solve the Big Bang riddles by plunging into the Planck time, $t_p = 10^{-42}$ s, before which the temperatures in the universe are so high that gravity, and therefore the evolution of the universe, becomes dominated by quantum effects. Perhaps the most challenging approach is quantum cosmology, an attempt to describe the universe with a wave function, hopefully subject to a Schrödinger-type equation. We shall not describe this side of the story. Instead we present a few ideas suggesting that before the Big Bang there may be another classical period in the life of the universe. In this previous incarnation one seeks solutions to the cosmological puzzles.

Historically the first such attempt was the bouncing universe. Closed universes expand to a maximum size and then recollapse, eventually reaching a Big Crunch. What if the crunch bounced into a bang? This cannot be achieved classically, but may be possible due to quantum effects, although this remains a speculation. In Figure 1.4 we plot the typical evolution of the scale factor $a$ in such models. The maximum size of the universe is related to its entropy. The second law of thermodynamics then requires that the bouncing universe gets bigger in each cycle.

A bouncing universe does not have a horizon. To see this let us ask the question: can a light ray in a closed universe ever get back to its starting

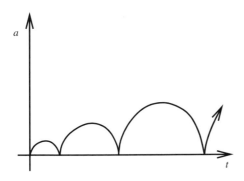

**Figure 1.4.** Closed models expand to a maximum size, then recollapse until they meet a Big Crunch. The bouncing universe model assumes that a crunch bounces into a new bang. This leads to a multi-cycle universe, in a succession of crunches and bangs, and to the prediction that there is a universe before our Big Bang.

point? Are there Magellanic photons in a spherical universe? The answer is yes: if a ray sets off at the Big Bang, it travels around the universe and gets back to the departure point at the Big Crunch. Hence, after the first cycle all points have been in causal contact. It is only if we are unaware of the cycles preceding our own that we may infer a horizon problem.

However, it turns out that even though we have solved the horizon problem, we have not solved the homogeneity problem. Ensuring causal contact between the whole observable universe allows for an equilibration mechanism to homogenise the whole universe, but such a mechanism must still be proposed and be efficient enough. No such mechanism seems to be present in bouncing universes.

A more modern way to explore life before the Big Bang was recently inspired by string theory. In string theory there are a number of duality symmetries, typically involving transforming big things into small things, and strong forces into weak forces. In the context of cosmology this is reflected in a scale-factor transformation of the form $a(t) \rightarrow a^{-1}(-t)$. This permits the extension of the history of the universe into times before the Big Bang, times $t < 0$. For such times the solution dual to the radiation post-Big Bang solution is inflationary expansion! The typical time evolution of the scale factor is described in Figure 1.5.

The pre-Big Bang is therefore really another way of getting inflation. But this time inflation occurs before the Big Bang. The pre-Big Bang scenario solves the horizon and flatness problems in the same way as inflation, but the timing is completely different.

Another aspect of the duality transformation assisting pre-Big Bang cosmology involves a field called the dilaton. The dilaton appears in all

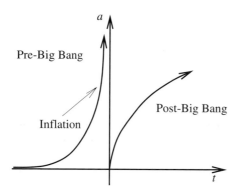

**Figure 1.5.** In pre-Big Bang cosmology the universe exists before the Big Bang. String theory symmetries allow one to project what we see after the Big Bang into what happens before the Big Bang. We find that there is inflation before the Big Bang. Inflation solves all the riddles of the Big Bang before the bang in this model.

attempts to derive low energy limits to string theory. It plays the role of a coupling constant for all interactions, or rather the couplings are related to this field. The duality transformation described above also requires a transformation upon the dilaton. This transformation requires that as time goes to $-\infty$ all coupling constants, describing the strength of interactions, go to zero. This means the pre-Big Bang universe emerges from an epoch in which the interactions were switched off.

The overall picture is that the universe starts with all interactions switched off. They then switch on and, in the process, inflation is also triggered, solving the riddles of the Big Bang. As the interactions become stronger and stronger, string theory effects become important, and lead the universe into the post-Big Bang stage. We do not know what happens in the Big Bang. However, we hope that the duality transformations assisting string theory will be enough to perform the mapping between these two stages in the life of the universe.

## 1.6 Quick-light

The special theory of relativity has dominated twentieth-century physics. More than the general theory, the special theory has become part of the fabric of physics. Special relativity has been successfully combined with quantum mechanics, to striking effect. Quantum field theory emerged from the union, with an array of spectacular predictions leading to modern particle physics. Good examples are the discovery of new particles and antiparticles, the electroweak theory and the prospect of unification (and the crucial idea of spontaneous symmetry breaking), as well as all sorts of high precision quantitative predictions concerning interactions and their cross-sections.

Central to special relativity is the idea that the speed of light $c$ is a constant. Regardless of the speed of the emitting or observing object, light moves at the same speed: about $300000\,\mathrm{km\,s^{-1}}$. Nothing can travel faster than light. The invariance of $c$ imposes a symmetry group on physics, called Lorentz symmetry. Only if space and time transform in a specific manner between different observers can the speed of light be the same for all of them. The implications of the Lorentz transformation are immensely popular. The (Lorentz) contraction of moving bodies, time dilation and the twin 'paradox' are now well known to everyone.

### 1.6.1 Varying speed of light

What if the speed of light were to change during the lifetime of the universe? 'Varying constant' theories have been proposed, starting with Dirac's idea of varying the gravitational constant $G$. In attempting to explain the constants of nature one should allow them to vary, and then see if a physical mechanism can be found which crystallises their values into fixed quantities. If so, we may hope that these values are the ones we observe. This project has not been terribly successful. But as a byproduct it has left us with great insights into what physics would be like if indeed the constants of nature were variables.

A good example is the Jordan–Brans–Dicke theory, in which the gravitational constant $G$ is a variable. In this theory the gravitational constant is the result of the matter content of the universe. As the cosmic density changes, $G$ changes as well. Such theories have led to interesting cosmological models, and attempts to solve Big Bang riddles with them have been made, albeit unsuccessfully. Another example is the theory proposed by Bekenstein, in which the electron charge $e$ is a field. String theories predict that all charges are in fact variable and related to a single field, the dilaton field.

Varying speed of light (VSL) is based on a similar exercise applied to $c$. In the simplest implementation of VSL, $c$ drops in a sharp phase transition in the early universe. Light was much faster in the early universe.

There is an element of criminal negligence in VSL. In VSL all observers at the same point, at the same time, but possibly moving relative to each other, see the same $c$. Again nothing can travel faster than light. However, Lorentz symmetry is broken. Once this happens we are in the dark. Lorentz symmetry has been the guiding principle used to set up all new theories in the twentieth century. If we discard it, what new guideline can we adopt?

We postulate a principle of minimal coupling. This means simply replacing $c$ by a field $c(t, \mathbf{x})$ wherever it occurs in selected laws. Such a minimal coupling principle guided the construction of other 'varying constant' theories. It ensures that nothing new happens when the 'varying constants' are kept fixed. It also ensures that minimal changes are introduced when 'varying constants' do vary.

Minimal coupling cannot be consistently applied in all laws. We decided to apply it to the field equations: in the case of gravity, to Einstein's equations. Curvature is not affected by VSL, and the way matter generates curvature is the same as before. In some loose sense we have general relativity without special relativity.

In the context of cosmology this means simply replacing $c$ by a variable $c(t)$ in Einstein's equations. This leads to the same space-time as before, and therefore we don't have superluminal expansion. VSL is not like inflation, thank you very much.

## 1.6.2 Quick-light years

It is immediately obvious that quick-light in the early universe solves the horizon problem. Look at Figure 1.6, in which we redraw Figure 1.3 assuming the speed of light was much larger in the early universe than it is nowadays, dropping to its current value in a sharp phase transition at $t = t_c$. We don't need to play tricks with expansion in order to establish causal contact over the whole observable universe. Even without expansion (so that $d_h = ct$ when $c$ is constant), the quick-light pervading the early universe would have been enough to connect the whole observable universe.

Suppose that the transition happened when the universe was one year old. The horizon was then one quick-light year across, easily bigger than 15 billion normal-light years, if quick-light is fast enough. If such a phase transition occurred at Planck time ($t_c = t_p$), then light would need to have been at least $10^{32}$ times faster than nowadays, to solve the horizon problem.

## 1.6.3 Another valley for flatness

Energy conservation appears in relativity as a consistency condition for Einstein's equations. This is spoiled if the speed of light is allowed to vary. We find that the same consistency conditions for Einstein's equations now

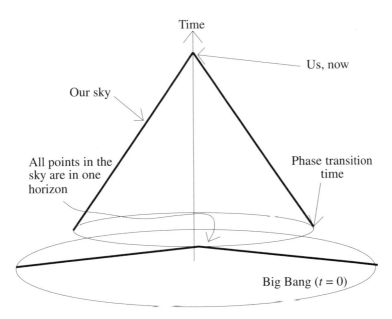

**Figure 1.6.** Diagram showing the horizon structure in a model in which at time $t_c$ the speed of light changed from $c^-$ to $c^+ \ll c^-$. Light travels at 45° after $t_c$ but it travels at a much smaller angle with the space axis before $t_c$. Hence it is possible for the horizon at $t_c$ to be much larger than the portion of the universe at $t_c$ intersecting our past light cone. All regions in our past have then always been in causal contact.

describe violations of energy conservation, proportional to the rate of change of the speed of light.

It is not surprising that energy conservation must be violated in such a theory. Conservation laws are the result of symmetries: that is the modern way to look at it. For instance, conservation of angular momentum is a consequence of isotropy, the symmetry according to which the universe looks the same in all directions. Energy conservation results from invariance under time translations: the laws of physics are the same at all times. This is clearly not the case if the speed of light changes: the laws of physics change fundamentally as the speed of light changes. Therefore there must indeed be violations of energy conservation if $c$ varies.

Lack of energy conservation pays its dividends. We can diagnose the geometry of the universe by looking at its energy density. The flat $K=0$ universe has a particular energy density, for a given expansion rate of the universe, called the critical energy density $\rho_c$. If the density is supercritical, $\rho > \rho_c$, then the universe is closed; if the density is subcritical, $\rho < \rho_c$, the universe is open. We find that if the speed of light decreases then matter is created if the universe is open ($K = -1$ and $\rho < \rho_c$), but disappears if the universe is closed ($K = 1$ and $\rho > \rho_c$). There is no creation or annihilation if the universe is flat ($K = 0$ and $\rho = \rho_c$). Hence VSL creates matter if we are subcritical, subtracts it if we are supercritical. Again we have produced a valley for flatness.

It can be shown that a drop in $c$ by 32 orders of magnitude at Planck time would provide sufficient fine tuning for a flat universe to be seen nowadays.

### 1.6.4 Exorcising the nothing

Finally VSL solves the cosmological constant problem. Einstein introduced $\Lambda$ into his equations as an extra geometrical term. However, the dynamical importance of $\Lambda$ can only be inferred when we reinterpret it as a fluid, with a density which remains constant under expansion. The density of this fluid is related to the speed of light. Indeed we find that $\rho_\Lambda \propto c^2$. We see how a drop in $c$ reduces the dynamical significance of $\Lambda$. If $c$ drops by more than 64 orders of magnitude at Planck time, then indeed $\rho_\Lambda \ll \rho$ nowadays. We have exorcised vacuum domination.

Celebrations of this triumph were interrupted by disturbing claims for observational evidence that the cosmic expansion is accelerating. This implies that $\Lambda$ is still with us, and is about to dominate the universe. We are about to enter a period of inflation!

This is horrifying. All galaxies will soon recede away from us so fast that we will not be able to see them. We will soon be confined to our galaxy-island, with nothing but the $\Lambda$-vacuum to keep us company, in cosmic loneliness. We will end up in an island universe, as Kant envisaged. Explaining why $\Lambda$ is only now about to dominate the universe is an outstanding challenge. Why now? Why not immediately after the Planck time? Why not never?

As this chapter goes to press one of the authors is suffering from insomnia due to this humiliating riddle.

### 1.7 An appraisal of current cosmology

In this review we provided a rather diluted version of a very technical field. We described how the Big Bang model converted cosmology into a successful science. We showed how its riddles have provided insights into theories of the very early universe, when the Big Bang must be replaced by something more fundamental. We described three classes of models. Inflation is now the leading theory, pre-Big Bang models are a popular tentative idea, while VSL theories are outright speculation. In the words of a distinguished Cambridge professor, 'VSL is a step back from relativity'.

In order to make this review more accessible, we highlighted the least technical aspects of the Big Bang riddles. This necessarily distorts the field. Perhaps the biggest riddle of all is the emergence of structure in a universe known to be very homogeneous at early times. All the above theories can answer this riddle, but in ways too technical for a light-hearted review like this one.

Nonetheless structure formation is really the testing ground where experiment may one day decide between all these ideas. We can measure the properties of galaxy clustering, and also the power spectrum in the cosmic microwave background (CMB) anisotropies. Theories of the early universe make very different predictions for these observations. A new generation of satellite CMB experiments, plus new galaxy surveys, leave us at a threshold. In the twenty-first century it could well be decided which, if any, of the above ideas is correct.

The most exciting possibility is of course that all the current ideas are proved wrong. For that reason, we believe that this is a bad time to adopt a dogmatic view in cosmology. Instead, we should try out as many new ideas as possible. Who knows, they may still all be wrong. We advocate promiscuity in science.

## 1.8 Further reading

Albrecht, A. & Magueijo, J. 1999 *Phys. Rev.* D **59** 043516.

Linde, A. 1990 *Inflation and quantum cosmology*. Academic Press.

Magueijo, J. & Baskerville, K. 1999 Millennium Issue *Phil Trans. R. Soc. Lond.* A **357** 3221–3236.

Zeldovich, Ya. & Novikov, I. 1983 *Relativistic astrophysics II*. University of Chicago Press.

# 2
# The origin of structure in the universe

## Juan García-Bellido

*Theoretical Physics Department C-XI, Autónoma University, Cantoblanco*
*28049 Madrid, Spain*

Cosmology is probably the most ancient body of knowledge, dating from as far back as the predictions of seasons by early civilisations. Yet, until recently, one could only answer to some of its more basic questions with an order of magnitude estimate. This poor state of affairs has dramatically changed in the last few years, thanks to important new data coming from precise measurements of a wide range of cosmological parameters that determine our standard cosmological model. We are entering a precision era in cosmology, and soon most of our observables will be measured to within a small percentage accuracy.

Our present understanding of the universe is based upon the successful hot Big Bang theory, which explains its evolution from the first fraction of a second to our present age, around 13 billion years later. This theory rests upon four strong pillars, a theoretical framework based on general relativity, as put forward by Albert Einstein and Alexander A. Friedmann in the 1920s, and three robust observational facts. First, the expansion of the universe, discovered by Edwin P. Hubble in the 1930s, as a recession of galaxies at a speed proportional to their distance from us. Second, the relative abundance of light elements, explained by George Gamow in the 1940s, mainly that of helium, deuterium and lithium, which were cooked from the nuclear reactions that took place at around a second to a few minutes after the Big Bang, when the universe was a hundred times hotter than the core of the Sun. Third, the cosmic microwave background (CMB), the afterglow of the Big Bang, discovered in 1965 by Arno A. Penzias and Robert W.

Wilson as a very isotropic blackbody radiation at a temperature of about 3 kelvin, emitted when the universe was cold enough to form neutral atoms, and photons decoupled from matter, approximately 300000 years after the Big Bang. Today, these observations are confirmed to within a small percentage accuracy, and have helped establish the hot Big Bang as the preferred model of the universe.

The Big Bang theory could not explain, however, the origin of matter and structure in the universe; that is, the origin of the matter–antimatter asymmetry, without which the universe today would be filled by a uniform radiation continuously expanding and cooling, with no traces of matter, and thus without the possibility to form gravitationally bound systems like galaxies, stars and planets that could sustain life. Moreover, the standard Big Bang theory assumes, but cannot explain, the origin of the extraordinary smoothness and flatness of the universe on the very large scales seen by the microwave background probes and the largest galaxy catalogues. It cannot explain the origin of the primordial density perturbations that gave rise to cosmic structures like galaxies, clusters and superclusters, via gravitational collapse; the quantity and nature of the dark matter that we believe holds the universe together; nor the origin of the Big Bang itself.

In the 1980s, a new idea, cosmological inflation, deeply rooted in fundamental particle physics, was put forward by Alan H. Guth, Andrei D. Linde and others, to address these fundamental questions. According to inflation, the early universe went through a period of exponential expansion, driven by the approximately constant energy density of a scalar field called the inflaton (see Figure 2.1). We know from general relativity that the density of matter determines the expansion of the universe, but a constant energy density acts in a very peculiar way: as a repulsive force that makes any two points in space separate at exponentially large speeds. (This does not violate the laws of causality because there is no information carried along in the expansion, it is simply the stretching of space-time.)

This superluminal expansion is capable of explaining the large-scale homogeneity of our observable universe and, in particular, why the microwave background looks so isotropic: regions separated today by more than 1° in the sky were, in fact, in causal contact before inflation, but were stretched to cosmological distances by the expansion (see Figure 2.2). Any inhomogeneities present before the tremendous expansion would be washed out.

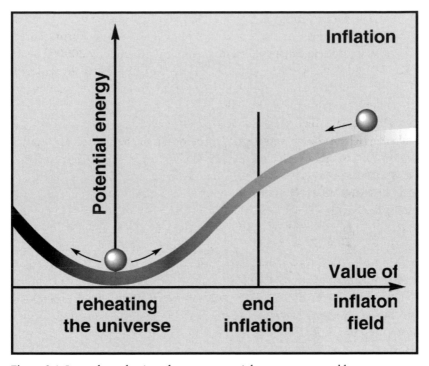

**Figure 2.1.** In modern physics, elementary particles are represented by quantum fields, which resemble the familiar electric, magnetic and gravitational fields. A field is simply a function of space and time whose quantum oscillations are interpreted as particles. For instance, the photon is the particle associated with the electromagnetic field. In our case, the inflaton field has, associated with it, a large potential energy density, which drives the exponential expansion during inflation. The inflaton field can be represented as a ball rolling down a hill. During inflation, the energy density is approximately constant, driving the tremendous expansion of the universe. When the ball starts to oscillate around the bottom of the hill, inflation ends and the inflaton energy decays into particles. In certain cases, the coherent oscillations of the inflaton could generate a resonant production of particles which soon thermalise, reheating the universe.

Moreover, in the usual Big Bang scenario a flat universe, one in which the gravitational attraction of matter is exactly balanced by the cosmic expansion, is unstable under perturbations: a small deviation from flatness is amplified and very quickly produces either an empty universe or a collapsed one. For the universe to be nearly flat today, as observations suggest,

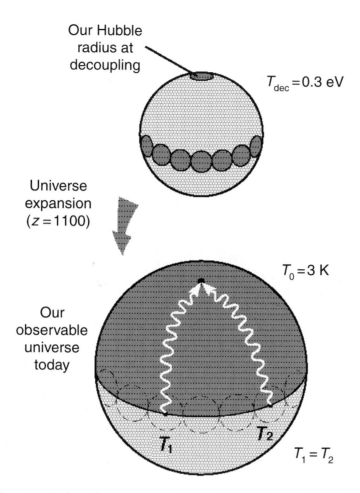

**Figure 2.2.** Perhaps the most acute problem of the Big Bang model is explaining the extraordinary homogeneity and isotropy of the microwave background. Information cannot travel faster than the speed of light, so the causal region (so-called horizon or Hubble radius) at the time of photon decoupling could not be larger than 300 000 light years across, or about 1° projected in the sky today. So why should regions that are separated by more than 1° in the sky have the same temperature, when the photons that come from those two distant regions could not have been in causal contact when they were emitted? This constitutes the so-called horizon problem, which is spectacularly solved by inflation: those regions were actually in causal contact before inflation, and therefore had the same properties, but were stretched to cosmological distances by the superluminal expansion.

it must have been extremely flat at nucleosynthesis, for example, deviations not exceeding more than one part in $10^{15}$. This extreme fine tuning of initial conditions was also solved by cosmological inflation. The exponential expansion made the radius of curvature of the universe so large that our observable patch of the universe today appears essentially flat, analogous (in three dimensions) to how the surface of a balloon appears flatter and flatter as we inflate it to enormous size. This is a crucial prediction of cosmological inflation that has been recently confirmed by observations of the microwave background anisotropies. Inflation is therefore an extremely elegant hypothesis that explains how a region much, much greater than our own observable universe could have become smooth and flat without recourse to *ad hoc* initial conditions.

If cosmological inflation made the universe so extremely flat and homogeneous, where did the galaxies and clusters of galaxies come from? One of the most astonishing predictions of inflation, one that was not even expected, is that quantum fluctuations of the inflaton field are also stretched by the exponential expansion and generate large-scale perturbations in the metric. Inflaton fluctuations are small wave packets of energy that, according to general relativity, modify the space-time fabric, creating a whole spectrum of curvature perturbations. The use of the word spectrum here is closely related to the case of light waves propagating in a medium: a spectrum characterises the amplitude of each given wavelength. In the case of inflation, the inflaton fluctuations induce waves in the space-time metric that can be decomposed into different wavelengths, all with approximately the same amplitude, that is, corresponding to a scale-invariant spectrum. These patterns of perturbations in the metric are like fingerprints that unequivocally characterise a period of inflation. When matter fell in the troughs of these waves, it created density perturbations that collapsed gravitationally to form galaxies, clusters and superclusters of galaxies, with a power spectrum that is also scale invariant. Such a type of spectrum was proposed in the early 1970s (before inflation) by Edward R. Harrison, and independently by the Russian cosmologist Yakov B. Zel'dovich, to explain the distribution of galaxies and clusters of galaxies on very large scales in our observable universe.

According to the Big Bang theory, the further away a galaxy is, the larger its recession velocity, and the larger the shift towards the red of the spectrum of light from that galaxy. Astronomers thus measure distances in units of redshift $z$. The furthest galaxies observed so far are at redshifts of

$z = 5$, or 12 billion light years from the Earth, whose light was emitted when the universe had only about 5 per cent of its present age. Only a few galaxies are known at those redshifts, but there are at present various catalogues like the CfA and APM galaxy catalogues, and more recently the IRAS PSCz (see Figure 2.3) and Las Campanas redshift surveys, that study the spatial distribution of hundreds of thousands of galaxies up to distances of a billion light years, or $z \lesssim 0.1$, that recede from us at speeds of tens of thousands of kilometres per second. These catalogues are telling us about the evolution of clusters of galaxies in the universe, and already put constraints on the theory of structure formation based on the gravitational collapse of the small inhomogeneities produced during inflation. From these

PSCz + BTP galaxies

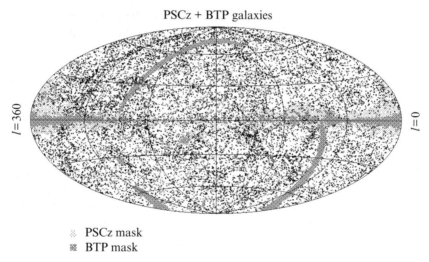

$l = 360$         $l = 0$

  PSCz mask
  BTP mask

**Figure 2.3.** Large galaxy catalogues allow cosmologists to study the spatial distribution of hundreds of thousands of galaxies up to distances of billions of light years. For instance, the IRAS Point Source Catalog redshift survey contains some 15 000 galaxies, covering over 83 per cent of the sky up to redshifts $z \lesssim 0.05$. We show here the projection of the galaxy distribution in galactic coordinates. It is clear that the distribution looks very homogeneous on very large scales, yet there is structure on both small and large scales. The filled-in regions indicate unobserved or obscured regions, in particular along the horizontal strip surrounding the galactic plane. Future galaxy surveys, like the Sloan Digital Sky Survey, will map a million galaxies up to redshifts $z \lesssim 0.5$ and will test the validity of the theory of structure formation, based on the gravitational collapse of the small inhomogeneities produced during inflation.

observations one can infer that most galaxies formed at redshifts of the order of 2–6; clusters of galaxies formed at redshifts of order 1, and super-clusters are forming now. That is, cosmic structure formed from the bottom up: from galaxies to clusters to superclusters, and not the other way around.

This fundamental difference is an indication of the type of matter that gave rise to structure. We know from primordial nucleosynthesis that all the baryons in the universe cannot account for the observed amount of matter, so there must be some extra matter (dark since we don't see it) to account for its gravitational pull. Whether it is relativistic (hot) or non-relativistic (cold) could be inferred from observations: relativistic particles tend to diffuse from one concentration of matter to another, thus transfer-ring energy among them and preventing the growth of structure on small scales. This is excluded by observations, so we conclude that most of the matter responsible for structure formation must be cold. How much there is, is a matter of debate at the moment. Some recent analyses suggest that there is not enough cold dark matter to reach the critical density required to make the universe flat. If we want to make sense of the present obser-vations, we must conclude that some other form of energy permeates the universe. In order to resolve this issue, even deeper galaxy redshift cata-logues are underway, looking at millions of galaxies, like the Sloan Digital Sky Survey (SDSS) and the Anglo-Australian two degree field (2dF) galaxy redshift survey, which are at this moment taking data, up to redshifts of $z \lesssim 0.5$, or several billion light years away, over a large region of the sky. These important observations will help astronomers determine the nature of the dark matter and test the validity of the models of structure formation.

However, if galaxies did indeed form from the gravitational collapse of density perturbations produced during inflation, one should also expect to see such ripples in the metric as temperature anisotropies in the cosmic microwave background, that is, minute deviations in the temperature of the blackbody spectrum when we look at different directions in the sky. Such anisotropies had been looked for ever since Penzias and Wilson's dis-covery of the CMB, but had eluded all detection until NASA's Cosmic Background Explorer (COBE) satellite found them in 1992. The reason why they took so long to be discovered was that they appear as perturbations in temperature of only one part in $10^5$. There is, in fact, a dipolar anisot-ropy of one part in $10^3$, in the direction of the Virgo cluster, but that is

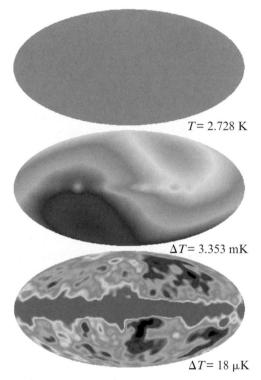

$T = 2.728$ K

$\Delta T = 3.353$ mK

$\Delta T = 18$ μK

**Figure 2.4.** The microwave background sky as seen by the Differential Microwave Radiometer on the Cosmic Background Explorer satellite. The upper panel shows the extraordinary homogeneity and isotropy of the universe: the microwave background has a uniform blackbody temperature of $T = 2.728$ K. The middle panel shows the dipole, corresponding to our relative motion with respect to the microwave background in the direction of the Virgo cluster. The lower panel shows the intrinsic CMB anisotropies, corresponding to the quadrupole and higher multipoles, at the level of one part in $10^5$. The horizontal bar corresponds to the microwave emission of our galaxy, which must be subtracted. These anisotropies contain crucial information about the cosmological parameters, and future satellites, like MAP and Planck, will measure them with unprecedented sensitivity and resolution, see Figure 2.5.

interpreted consistently as our relative motion with respect to the microwave background due to the local distribution of mass, which attracts us gravitationally towards the Virgo cluster. When subtracted, we are left with a whole spectrum of anisotropies in the higher multipoles (quadrupole, octupole, etc.) (see Figure 2.4). Soon after COBE, other groups quickly confirmed the detection of temperature anisotropies at around 30μK and above, at higher multipole numbers or smaller angular scales.

There are at this moment dozens of ground and balloon-borne experiments studying the anisotropies in the microwave background with angular resolutions from 1° to a few arc minutes in the sky. The physics of the CMB anisotropies is relatively simple: photons scatter off charged particles (protons and electrons), and carry energy, so they feel the gravitational potential associated with the perturbations imprinted in the metric during inflation. An overdensity of baryons (protons and neutrons) does not collapse under the effect of gravity until it enters the causal Hubble radius. The perturbation continues to grow until radiation pressure opposes gravity and sets up acoustic oscillations in the plasma, very similar to sound waves. Since overdensities of the same size will enter the Hubble radius at the same time, they will oscillate in phase. Moreover, since photons scatter off these baryons, the acoustic oscillations occur also in the photon field and induce a pattern of peaks in the temperature anisotropies in the sky, at different angular scales (see Figure 2.5). The larger the amount of baryons, the higher the peaks. The first peak in the photon distribution corresponds to overdensities that have undergone half an oscillation, that is, a compression, and appear at a scale associated with the size of the horizon at last scattering (when the photons decoupled) or about 1° in the sky. Other peaks occur at harmonics of this, corresponding to smaller angular scales. Since the amplitude and position of the primary and secondary peaks are directly determined by the sound speed (and, hence, the equation of state) and by the geometry and expansion of the universe, they can be used as a powerful test of the density of baryons and dark matter, and other cosmological parameters. For instance, recent measurements of the position of the first acoustic peak have confirmed that we live in a spatially flat universe, in agreement with the predictions of inflation.

By looking at these patterns in the anisotropies of the microwave background, cosmologists can also determine the primordial spectrum of density perturbations produced during inflation. It turns out that the observed temperature anisotropies are compatible with a scale-invariant spectrum, as predicted by inflation. This is remarkable, and gives very strong support to the idea that inflation may indeed be responsible for both the CMB anisotropies and the large-scale structure of the universe. Different models of inflation have different specific predictions for the fine details associated with the spectrum generated during inflation. It is these minute differences that will allow cosmologists to differentiate between alternative models of inflation and discard those that do not agree with

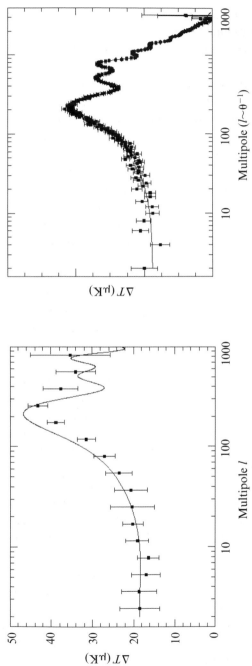

**Figure 2.5.** There are at present dozens of ground and balloon-borne experiments looking at the microwave background temperature anisotropies ($\Delta T$, in units of microkelvin) with angular resolutions from $10°$ to a few arc minutes in the sky, corresponding to multipole numbers $l = 2$–3000. Present observations suggest the existence of a peak in the angular distribution, as predicted by inflation (left panel, c. 1999). The theoretical curve (thin line) illustrates a particular model which fits the data. Future satellites, like MAP and Planck, will be able to measure, in a few years' time, the microwave background anisotropies with unprecedented resolution and sensitivity (right panel, c. 2008), and thus determine the cosmological parameters of the standard model of cosmology with extraordinary accuracy, see Table 2.1. (Figure by D. Scott (1999).)

observations. However, most importantly perhaps, the pattern of anisotropies predicted by inflation is completely different from those predicted by alternative models of structure formation, like cosmic defects: strings, vortices, textures, etc. These are complicated networks of energy density concentrations left over from an early universe phase transition, analogous to the defects formed in the laboratory in certain kinds of liquid crystals when they go through a phase transition. Cosmological defects have spectral properties very different from those generated by inflation. That is why it is so important to launch more sensitive instruments, and with better angular resolution, to determine the properties of the CMB anisotropies, and discard alternative theories of structure formation.

In the next few years two new satellites – the Microwave Anisotropy Probe (MAP), to be launched by NASA in the year 2001, and the Planck Surveyor, due to be launched by the European Space Agency in 2007 – will measure those temperature anisotropies with 100 times better angular resolution and 10 times better sensitivity than COBE, and thus allow cosmologists to determine the parameters of the standard cosmological model with unprecedented accuracy. What makes the microwave background observations particularly powerful is the absence of large systematic errors that plague other cosmological measurements. As discussed above, the physics of the microwave background is relatively simple compared to, say, the physics of supernova explosions, and computations can be done consistently within perturbation theory. Thus, most of the systematic errors are theoretical in nature, due to our ignorance about the primordial spectrum of metric perturbations from inflation. There is a great effort at the moment in trying to cover a large region in the parameter space of models of inflation, to ensure that we have considered all possible alternatives, like isocurvature or pressure perturbations, non-scale invariant or tilted spectra and non-Gaussian density perturbations. In particular, inflation also predicts a spectrum of gravitational waves. Their amplitude is directly proportional to the total energy density during inflation, and thus its detection would immediately tell us about the energy scale (and, therefore, the epoch in the early universe) at which inflation occurred. If the period of inflation responsible for the observed CMB anisotropies is associated with the grand unification scale, when the strong and electroweak interactions are supposed to unify, then there is a chance that we might see the effect of gravitational waves in the future satellite measurements, especially from the analysis of photon polarisation in the microwave background maps.

In the quest for the parameters of the standard cosmological model, various groups are searching for distant astrophysical objects that can serve as standard candles to determine the distance to an object from its observed apparent luminosity. A candidate that has recently been exploited with great success is a certain type of supernova explosion at large redshifts. These are stars at the end of their life cycle that become unstable and violently explode in a natural thermonuclear explosion that out-shines their progenitor galaxy. The intensity of the distant flash varies in time; it takes about three weeks to reach its maximum brightness and then it declines over a period of months. Although the maximum luminosity varies from one supernova to another, depending on their original mass, their environment, etc., there is a pattern: brighter explosions last longer than fainter ones. By studying the light curves of a reasonably large statistical sample, cosmologists from two competing groups, the Supernova Cosmology Project and the High-redshift Supernova Project, are confident that they can use this type of supernova as a standard candle. Since the light coming from some of these rare explosions has travelled for a large fraction of the size of the universe, one expects to be able to infer, from their distribution, the spatial curvature and the rate of expansion of the universe (see Figure 2.6). One of the surprises revealed by these observations is that the universe appears to be accelerating instead of decelerating, as was expected from the general attraction of matter; something seems to be acting as a repulsive force on very large scales. The most natural explanation for this is the existence of a cosmological constant, a diffuse vacuum energy that permeates all space and gives the universe an acceleration that tends to separate gravitationally bound systems from each other. The origin of such a vacuum energy is one of the biggests problems of modern physics. Its observed value is 120 orders of magnitude smaller than predicted by quantum physics. If confirmed, it will pose a real challenge to theoretical physics, one that may affect its most basic foundations. However, it is still premature to conclude that this is indeed the case, because of possibly large systematic errors inherent to most cosmological measurements, given the impossibility of performing experiments under similar circumstances in the laboratory.

Cosmological inflation may be responsible for the metric perturbations that later gave rise to the large-scale structures we see in the universe, but where did all the matter in the universe come from? Why isn't all the energy in photons, which would have inevitably redshifted away in a cold

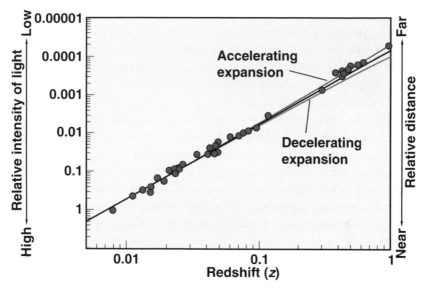

**Figure 2.6.** The Hubble diagram shows the luminosity distance to an object (measured by the intensity of light from that object) as a function of redshift. The proportionality constant depends on the Hubble rate of expansion and the matter content of the universe. Cosmologists use calibrated 'standard candles' as distant beacons for the determination of the universe expansion. Recent observations of high redshift supernovae (dots) show that the universe deviates slightly but significantly from the Einstein–de Sitter model (lower line), a flat universe with no cosmological constant, which would be decelerating due to the attraction of matter. These observations indicate that there is only 30 per cent of the matter necessary to make it flat (middle line), and, therefore, it decelerates more slowly than predicted. The measurements even suggest that the universe is accelerating (upper line), as if due to a non-zero cosmological constant.

universe devoid of life? How did we end up being matter dominated? Everything we see in the universe, from planets and stars, to galaxies and clusters of galaxies, is made out of matter, so where did the antimatter in the universe go? Is this the result of an accident, a happy chance occurrence during the evolution of the universe, or is it an inevitable consequence of some asymmetry in the laws of nature? Theorists believe that the excess of matter over antimatter comes from fundamental differences in their interactions soon after the end of inflation.

Inflation is an extremely efficient mechanism in diluting any particle species or fluctuations. At the end of inflation, the universe is empty and extremely cold, dominated by the homogeneous coherent mode of the inflaton. Its potential energy density is converted into particles, as the inflaton field oscillates coherently around the minimum of its potential (see Figure 2.1). These particles are initially very far from equilibrium, but they interact strongly among themselves and soon reach thermal equilibrium at a very large temperature. From there on, the universe expands iso-entropically, cooling down as it expands, in the way described by the standard hot Big Bang model. Thus the origin of the Big Bang itself, and the matter and energy we observe in the universe today, can be traced back to the epoch in which the inflaton energy density decayed into particles. Such a process is called reheating of the universe. Recent developments in the theory of reheating suggest that the decay of the inflaton energy could be explosive due to the coherent oscillations of the inflaton, which induce its stimulated decay. The result is a resonant production of particles in just a few inflaton oscillations, an effect very similar to the stimulated emission of a laser beam of photons. The number of particles produced this way is exponentially large, which may explain the extraordinarily large entropy, of order $10^{89}$ particles, in our observable patch of the universe today. However, the inflaton is supposed to be a neutral scalar field, and thus its interactions cannot differentiate between particles and antiparticles. How did we end up with more matter than antimatter? The study of this cosmological asymmetry goes by the name of baryogenesis since baryons (mainly protons and neutrons) are the fundamental constituents of matter in planets, stars and galaxies in the universe today. Therefore, what are the physical conditions for baryogenesis?

Everything we know about the properties of elementary particles is included in the standard model of particle physics. It describes more than 100 observed particles and their interactions in terms of a few fundamental constituents: six quarks and six leptons, and their antiparticles. The standard model describes three types of interactions: the electromagnetic force, the strong and the weak nuclear forces. These forces are transmitted by the corresponding particles: the photon, the gluon and the W and Z bosons. The theory also requires a scalar particle, the Higgs particle, responsible for the masses of quarks and leptons and the breaking of the electroweak symmetry at an energy scale 1000 times the mass of the proton. The Higgs is believed to lie behind most of the mysteries of the standard model, including the asymmetry between matter and antimatter.

Symmetries are fundamental properties of any physical theory. A theory is symmetric under certain symmetry operation, like reflection, if its laws apply equally well after such an operation is performed on part of the physical system. An important example is the operation called parity reversal, denoted by $P$. It produces a mirror reflection of an object and rotates it 180° about an axis perpendicular to the mirror. A theory has $P$ symmetry if the laws of physics are the same in the real and the parity-reversed world. Particles such as leptons and quarks can be classified as right- or left-handed depending on the sense of their internal rotation, or spin, around their direction of motion. If $P$ symmetry holds, right-handed particles behave exactly like left-handed ones. The laws of electrodynamics and the strong interactions are the same in a parity-reflected universe. But, as Chien-Shiung Wu discovered in 1957, the weak interaction acts very differently on particles with different handedness: only left-handed particles can decay by means of the weak interaction, not right-handed ones. Moreover, as far as we know, there are no right-handed neutrinos, only left-handed. So the weak force maximally violates $P$.

Another basic symmetry of nature is charge conjugation, denoted by $C$. This operation changes the quantum numbers of every particle into those of its antiparticle. Charge symmetry is also violated by the weak interactions: antineutrinos are not left-handed, only right-handed. Combining $C$ and $P$, one gets the charge–parity symmetry $CP$, which turns all particles into their antiparticles and also reverses their handedness: left-handed neutrinos become right-handed antineutrinos. Although charge and parity symmetry are individually broken by the weak interaction, their combination was expected to be conserved. However, in 1964, a groundbreaking experiment by James Cronin, Val Finch and Renè Turlay at Brookhaven National Laboratory showed that $CP$ was in fact violated to one part in 1000. It was hard to see why $CP$ symmetry should be broken at all, and even more difficult to understand why the breaking is so small. Soon after, in 1972, Makoto Kobayashi and Toshihide Maskawa showed that $CP$ could be violated within the standard model if three or more generations of quarks existed, because of $CP$ non-conserving phases that could not be rotated away. Only two generations of particles were known at the time, but, in 1975, Martin L. Perl and collaborators discovered the tau lepton at the Stanford Linear Accelerator Center (SLAC), the first ingredient of the third generation. Only recently the last quark in the family, the top quark, was discovered at Fermilab.

But how does this picture fit in the evolution of the universe? In 1967, the Russian physicist Andrei Sakharov pointed out the three necessary conditions for the baryon asymmetry of the universe to develop. First, we need interactions that do not conserve baryon number $B$, otherwise no asymmetry could be produced in the first place. Second, $C$ and $CP$ symmetry must be violated, in order to differentiate between matter and antimatter, otherwise $B$ non-conserving interactions would produce baryons and antibaryons at the same rate, thus maintaining zero net baryon number. Third, these processes should occur out of thermal equilibrium, otherwise particles and antiparticles, which have the same mass, would have equal number densities and would be produced at the same rate. The standard model is baryon symmetric at the classical level, but violates $B$ at the quantum level, through the chiral anomaly. Electroweak interactions violate $C$ and $CP$, but the magnitude of the latter is clearly insufficient to account for the observed baryon asymmetry. This failure suggests that there must be other sources of $CP$ violation in nature, and thus the standard model of particle physics is probably incomplete.

One of the most popular extensions of the standard model includes a new symmetry called supersymmetry, which relates bosons (particles that mediate interactions) with fermions (the constituents of matter). Those extensions generically predict other sources of $CP$ violation coming from new interactions at scales above 1000 times the mass of the proton. Such scales will soon be explored by particle colliders like the Large Hadron Collider (LHC) at CERN (the European Centre for Particle Physics) and by the Tevatron at Fermilab. The mechanism for baryon production in the early universe in these models relies on the strength of the electroweak phase transition, as the universe cooled and the symmetry was broken. Only for strongly first-order phase transitions is the universe sufficiently far from equilibrium to produce enough baryon asymmetry. Unfortunately, the phase transition in these models is typically too weak to account for the observed asymmetry, so some other mechanism is needed. If reheating after inflation occurred in an explosive way, via the resonant production of particles from the inflaton decay, as recent theoretical developments suggest, then the universe has actually gone through a very nonlinear, non-perturbative and very-far-from-equilibrium stage, before thermalising via particle interactions. Electroweak baryogenesis could have taken place during that epoch, soon after the end of low-energy inflation at the electroweak scale. Such models can be constructed but require a specially flat direction (a very small mass for the inflaton) during inflation. After inflation, the inflaton acquires a large mass from its interaction with the Higgs field, and could be seen in future particle colliders.

The crucial ingredient of departure from equilibrium, necessary for the excess production of baryons over antibaryons, is strongly present in this new scenario of baryogenesis, as the universe develops from a zero-temperature and zero-entropy state, at the end of inflation, to a thermal state with exponentially large numbers of particles, the origin of the standard hot Big Bang. If, during this stage, fundamental or effective interactions that are $B$, $C$ and $CP$ violating were fast enough compared to the rate of expansion, the universe could have ended up with the observed baryon asymmetry of one baryon per $10^9$ photons today, as deduced from observations of the light element abundances. Recent calculations suggest that, indeed, the required asymmetry could be produced as long as some new physics, just above the electroweak symmetry-breaking scale, induces an effective $CP$ violating interaction. These new phenomena necessarily involve an interaction between the Higgs particle, responsible for the electroweak symmetry breaking, and the inflaton field, responsible for the period of cosmological inflation. Therefore, for this scenario to work, it is expected that both the Higgs and the inflaton particles will be discovered at the future particle physics colliders like the LHC and the Future Linear Collider, to be built in the next few years. If confirmed, such a new scenario of baryogenesis would represent a leap forward in our understanding of the universe from the unifying theory of inflationary cosmology. Furthermore, it would bring inflation down to a scale where present or future particle physics experiments would be able to explore it quite thoroughly. Cosmological inflation thus enters the realm of testable high-energy particle physics.

We have entered a new era in cosmology, where a host of high-precision measurements are already posing challenges to our understanding of the universe: the density of ordinary matter and the total amount of energy in the universe; the microwave background anisotropies on a fine-scale resolution; primordial deuterium abundance from quasar absorption lines; the acceleration parameter of the universe from high-redshift supernovae observations; the rate of expansion from gravitational lensing; large-scale structure measurements of the distribution of galaxies and their evolution; and many more, which already put constraints on the parameter space of cosmological models (see Figure 2.7). However, these are only the forerunners of the precision era in cosmology that will

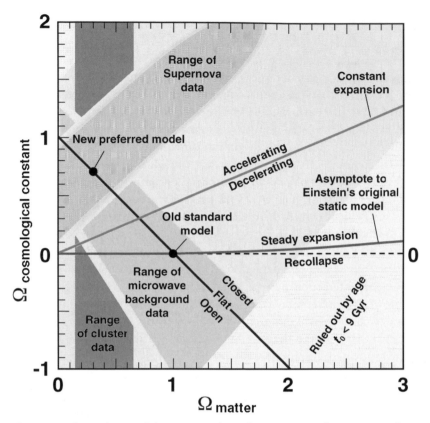

**Figure 2.7.** The evolution of the universe depends on two crucial parameters: the average density of matter (horizontal axis) and the energy density in the cosmological constant (vertical axis). Their values produce three very different effects. First, their sum gives the total cosmic energy content and determines the geometry of space-time, whether spatially flat, open or closed. Second, their difference characterises the relative strength of expansion and gravity, and determines how the expansion rate changes with time, whether accelerating or decelerating. The third, a balance between the two densities, determines the fate of the universe, whether it will expand forever or recollapse. These three effects have been probed by recent observations, from large-scale structure (cluster data), temperature anisotropies (microwave background data) and the universe expansion (supernova data). Surprisingly enough, at present, all observations seem to lie within a narrow region of parameter space. The Einstein–de Sitter model is no longer our preferred model. The best model today is a flat model with a third of the energy density in non-relativistic matter and two-thirds as a cosmological constant.

Table 2.1. *The parameters of the standard cosmological model*

| Physical quantity | Symbol | Present range | MAP (%) | Planck (%) |
|---|---|---|---|---|
| Luminous matter | $\Omega_{\text{lum}} h^2$ | 0.001–0.005 | — | — |
| Baryonic matter | $\Omega_{\text{B}} h^2$ | 0.01–0.03 | 5 | 0.6 |
| Cold dark matter | $\Omega_{\text{M}} h^2$ | 0.2–1.0 | 10 | 0.6 |
| Hot dark matter | $\Omega_{\nu} h^2$ | 0–0.3 | 5 | 2 |
| Cosmological constant | $\Omega_{\Lambda} h^2$ | 0–0.8 | 8 | 0.5 |
| Spatial curvature | $\Omega_0 h^2$ | 0.2–1.5 | 4 | 0.7 |
| Rate of expansion | $h$ | 0.4–0.8 | 11 | 2 |
| Age of the universe | $t_0$ | 11–17 Gyr | 10 | 2 |
| Spectral amplitude | $Q_{\text{rms}}$ | 20–30 μK | 0.5 | 0.1 |
| Spectral tilt | $n_{\text{S}}$ | 0.5–1.5 | 3 | 0.5 |
| Tensor–scalar ratio | $r_{\text{ts}}$ | 0–1.0 | 25 | 10 |
| Reionisation parameter | $\tau$ | 0.01–1.0 | 20 | 15 |

*Notes:*

The standard model of cosmology is characterised by around 12 different parameters, needed to describe the background space-time, the matter and energy content, and the primordial spectrum of density perturbations. We include here the present range of the most relevant parameters, and the percentage error with which the microwave background probes MAP and Planck will be able to determine them in the near future. (The rate of expansion is written in units of $H = 100h$ km s$^{-1}$ Mpc$^{-1}$.)

dominate the new millennium, and will make cosmology a phenomenological science.

In the next few years we will have a very large set of high-quality observations that will test inflation and the cold dark matter theory of structure formation, and determine most of the 12 or more parameters of the standard cosmological model to within a small percentage accuracy (see Table 2.1). It may seem that with such a large number of parameters one can fit almost anything. However, that is not the case when there is enough quantity and quality of data. An illustrative example is the standard model of particle physics, with around 21 parameters and a host of precise measurements from particle accelerators all over the world. This model is, nowadays, rigorously tested, and its parameters measured to a precision better than 1% in some cases. It is clear that high-precision measurements will make the standard model of cosmology as robust as that of particle physics. We are truly living in the Golden Age of Cosmology. With the advent of better and larger precision experiments, cosmology is becoming a mature science, where speculation has given way to phenomenology; but there is still a lot to do.

We have yet to answer many fundamental questions in this emerging picture of cosmology. For instance, we ignore the nature of the inflaton field; is it some new fundamental scalar field in the electroweak symmetry-breaking sector, or is it just some effective description of a more fundamental high-energy interaction? Hopefully, in the near future, experiments in particle physics might give us a clue to its nature. Inflation had its original inspiration in the Higgs field, the scalar field supposed to be responsible for the masses of elementary particles (quarks and leptons) and the breaking of the electroweak symmetry. Such a field has not been found yet, and its discovery at the future particle colliders would help us to understand one of the truly fundamental problems in physics, the origin of masses. If the experiments discover something completely new and unexpected, it would automatically affect inflation at a fundamental level.

One of the most difficult challenges that the new cosmology will have to face is understanding the origin of the cosmological constant, if indeed it is confirmed by independent sets of observations. Ever since Einstein introduced it as a way to counteract gravitational attraction, it has haunted cosmologists and particle physicists for decades. We still do not have a mechanism to explain its extraordinarily small value, 120 orders of magnitude below what is predicted by quantum physics. For several decades there has been the reasonable speculation that this fundamental problem may be related to the quantisation of gravity. General relativity is a classical theory of space-time, and it has proved particularly difficult to construct a consistent quantum theory of gravity, since it involves fundamental issues like causality and the nature of space-time itself. We can speculate that perhaps general relativity is not the correct description of gravity on the very largest scales. In fact, it is only in the last few billion years that the observable universe has become large enough for these global effects to be noticeable. In its infancy, the universe was much smaller than it is now, and, presumably, general relativity gave a correct description of its evolution, as confirmed by the successes of the standard Big Bang theory. As it expanded, larger and larger regions were encompassed, and, therefore, deviations from general relativity could slowly become important. It may well be that the recent determination of a cosmological constant from observations of supernovae at high redshifts is hinting at a fundamental misunderstanding of gravity at the quantum level, whose only manifestation today may be on the very large scales. If this were indeed the case, we should expect that the new generation of precise cosmological observations will not only affect our model of the universe but also a more fundamental description of nature.

## 2.1 Further reading

Guth, A. H. 1997 *The inflationary universe*, Reading: Perseus Books.

Hogan, C. J., Kirshner, R. P. & Suntzeff, N. B. 1999 Surveying space-time with supernovae, *Scientific Am.* January, pp. 46–51.

Krauss, L. M. 1999 Cosmological antigravity, *Sci. Am.* January, pp. 53–59.

Linde, A. D. 1994 The self-reproducing inflationary universe, *Sci. Am.* November, pp. 32–39.

Peebles, P. J. E., Schramm, D. N., Turner, E. L. & Kron, R. G. 1994 The evolution of the universe, *Sci. Am.* October, pp. 53–57.

Quinn, H. R. & Witherell, M. S. 1998 The asymmetry between matter and antimatter, *Sci. Am.* October, pp. 76–81.

A more complete list of references can be found in the special Millennium Issue, Part 1, García-Bellido, J. (1999) *Phil. Trans. R. Soc. Lond.* A **357**, 3237–3257.

# 3
# The dark side of the universe

## Ben Moore

*Department of Physics, University of Durham, Durham DH1 3LE, UK*

For over 50 years astronomers have been making a careful inventory of the material that makes up our universe. These observational studies have revealed the presence of a mysterious component of material that does not radiate light, yet amounts to more than 90 per cent of our known universe. Even accounting for this extra 'dark matter', the total mass is well below that required to prevent the universe from expanding forever. The determination of the nature of dark matter is recognised as one of the most fundamental unsolved problems in modern cosmology – revealing its nature would allow us to understand how galaxies form, to construct the history of the universe and also to predict its fate. This material plays a crucial role in the formation and evolution of structure in the universe and it is unlikely that galaxies and stars, or perhaps even life itself, would have formed without its presence.

The first evidence for dark matter was found by the astronomer Fritz Zwicky in 1933, who was puzzled by the large velocities of individual galaxies in the massive Coma Berenices cluster. Zwicky argued that the entire collection of galaxies would rapidly fly apart, unless there was a considerable amount of additional 'invisible' material present to bind the system together. Indications that individual galaxies contained more mass than could be accounted for in the visible component began accumulating at the same time as Zwicky's findings. However, it wasn't until the 1970s that it became generally accepted that galaxies contain a substantial component of mass in some unknown form and that this missing mass in galaxies is somehow related to the missing mass needed to hold galaxy clusters together.

Since these early investigations a host of observational data has accumulated, from the internal structure of galaxies to their large-scale clustering properties. Recently we have seen new controversial results, such as the galactic micro-lensing studies towards the Magellanic Clouds (the interpretation of which is being debated) and the resolution of old controversies such as the rejection of dark matter located in the galactic disc and solar neighbourhood. Theoretical modelling and interpretation of the observational data has also improved, in part due to the high quality of the numerical studies of the formation of structure in different cosmological models. The next decade is anticipated to be a golden era for cosmological research with the launch of several space-based missions designed to answer fundamental and outstanding questions about our universe. These multi-national projects will make precision observations of the universe and accurately measure the fundamental parameters that govern our cosmological world model. If all goes to plan they will accurately determine the amount and distribution of dark matter. However, we may still not know the identity of the dark matter until it is detected directly, whether through observations or within a laboratory.

## 3.1 Observational evidence

### 3.1.1 Dark matter on the smallest scales

Before we can look towards potential breakthroughs it is instructive to look back at the history of discovery and to review what we already know about dark matter. The first compelling argument for additional mass associated with individual galaxies was made by Kahn and Waljter in 1959, who used the dynamics of the Milky Way–Andromeda binary system to estimate its total mass. The Andromeda nebula is on a collision course with our own galaxy, currently moving at $120\,\mathrm{km\ s^{-1}}$ toward us; it will collide with the Milky Way in a few billion years time. The relative motion has been generated by the mutual gravitational attraction of the two galaxies over a timescale equal to the age of the universe – approximately 15 billion years. Solving the equation of motion for the system is straightforward and gives a total mass of the Local Group that is four thousand billion times the mass of the Sun, of which just one per cent can be accounted for by the visible stars.

Scientific discoveries sometimes occur from single creative advances in observational or theoretical understanding; in many cases it is the gradual accumulation of observations that leads to the acceptance of the data and new ideas. The acceptance of these data was delayed by a strong prejudice against the notion that practically nothing was known about the content of the universe. However, confirmation has since been found in many astrophysical systems using many different independent techniques. Galactic rotation curves now provide the strongest evidence for dark matter associated with individual galaxies. Most of the galaxies in the universe are spirals, like our own Milky Way, wherein the stars and gas move on circular orbits confined to a thin disc-like plane, therefore observing the rotational velocity at a given position gives a direct estimate of the internal mass. The contribution of the visible component (predominantly stars and gas) is insufficient to account for the observed speed, revealing the need for an additional dark component of matter (see Figure 3.1). At the edge of the galactic disc there is roughly ten times as much dark matter as baryons, although the total mass of the dark component is unknown since the stars or gas become too faint or diffuse to observe. The total extent of the dark matter surrounding galaxies is one of the key questions that future observations should help to solve.

One of the key observational facts that we have learned about the dark matter halos of many galaxies is that they have a unique structure with a large inner region that has close to a uniform density of dark matter. This causes the discs of spiral galaxies to rotate more slowly as one looks at smaller radii, finally reaching zero velocity at their centres. At larger radii the rotation curves flatten out to a constant rotation speed revealing that the total mass is increasing in proportion to distance from the centre (e.g. Figure 3.2). A declining rotation curve, which would signify the outer limit of dark matter halos, has never been observed. All of these properties are fundamental and must be reproduced by any prospective cosmological model for structure formation and dark matter candidate. One piece of observational evidence that has eluded astronomers so far is the measurement of the total extent of dark matter halos surrounding galaxies. We shall see later that the most successful model for structure formation predicts that the dark matter extends to beyond 300000 parsecs for a galaxy like the Milky Way (one parsec is roughly three light years). Confirming this prediction is extremely difficult since the visible disc only probes the central few percent of the expected mass distribution.

**Figure 3.1.** A spiral galaxy embedded within its halo of dark matter. This is an impression of the dark matter that literally fills the space within and around the galaxy, although it is thought to extend to ten times the distance drawn here.

If a system of objects, such as stars or galaxies, are gravitationally bound together and are in equilibrium then the total kinetic energy and potential energy must exactly balance. Therefore by measuring the mean velocity and radius of the system the total mass can be inferred. Unfortunately, galaxy halos contain very few visible objects that can be used to measure these quantities – just 12 satellite galaxies surround the Milky Way and the mass may extend much further than traced by these objects. Co-adding the satellite distributions of a large sample of bright spirals is one way of improving the statistics and allows the properties of the 'average halo' to be quantified. This work has provided compelling evidence that dark matter does indeed extend beyond 100000 parsecs; however, the same data has also yielded some very puzzling results that

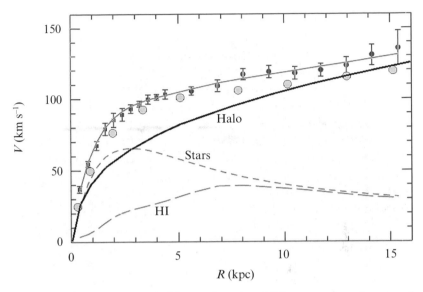

**Figure 3.2.** The rotation speed of the nearby galaxy M33 measured as a function of distance from the centre. The points with error bars are the observational data and the dashed curves show the contribution to the total rotation speed of the stars and gas. The sum of these components can only account for 10 per cent of the maximum rotational velocity, which can only be explained by adding an additional component of dark mass as indicated by the thick solid curve. The large shaded circles are the predicted rotation speed for a cold dark matter halo that should be compared to the observed halo contribution (the thick solid curve).

have yet to be explained. Intriguingly, satellites appear to avoid orbiting within the plane defined by the parent galaxy's disc and the total amount of dark matter surrounding each galaxy appears to be independent of its luminosity.

There is a promising new technique that may unravel some of these mysteries. Massive objects act as giant gravitational lenses that deflect light, causing detectable image distortions and magnification of background galaxies. The basic effect was predicted by general relativity and confirmed by observations in 1919. The amount of image distortion and magnification depends on the foreground mass distribution. By observing distant galaxies that lie behind a massive galaxy cluster, an independent measurement of the mass of the cluster can be made. This technique has been successfully applied to many different systems of which the example shown later in Figure 3.3 is one of the most spectacular. This effect is much weaker for galaxy halos that distort the images of background galaxies by less than one per cent, therefore the properties of any given dark matter halo cannot be inferred. Nevertheless, co-adding the data for a large number of similar galaxies has verified that the dark matter does indeed extend beyond 100 000 parsecs. Although observational data are still being accumulated, the total extent and structure of the dark matter halos that encompass galaxies are poorly understood. Gravitational lensing surveys will help resolve this problem and we await data of sufficient quality, such as will be obtained by the Next Generation Space Telescope or a dedicated 'dark matter lensing telescope'.

The dynamics of small stellar associations that orbit within our galactic halo provide some fascinating insights into the nature of the missing mass. The smallest galaxies that contain dark matter are the dwarf spheroidals. These tiny galaxies have a similar luminosity to globular clusters containing about a million stars, but they have scale lengths of just a few hundred parsecs, far larger than that of the compact star clusters but tiny compared to the Milky Way. When a handful of stellar motions were first measured the results were quite unexpected. Instead of moving randomly at typical velocities of about one kilometre per second, the stars were found to be moving ten times faster. Since the mass is proportional to the velocity squared, these galaxies contained 100 times as much dark material as visible stars!

Collecting spectra is much easier today with large multi-optic fibre systems on modern telescopes that can measure the velocities of many hundreds of stars simultaneously. We now know that dwarf spheroidals are completely dominated by dark matter at all radii. These galaxies have the highest dark matter content of any known galactic system and are also the smallest objects within which dark matter has been detected. Intriguingly, globular clusters have similar numbers of stars but are 10–100 times smaller and do not appear to contain any dark matter. Tidal streams of stars can be observed escaping from these systems, torn away by the larger gravitational field of the Milky Way that removes the most loosely bound stars. If globular clusters contained dark matter, or had extended halos of dark matter, then the stars would remain bound and these features would not be observed. Recent observations of stars escaping from the Carina dwarf galaxy can be used to carry out similar studies and to infer the extent of its dark matter halo.

### 3.1.2 Clusters of galaxies and beyond

Clusters of galaxies are the most massive structures that exist that are in gravitational equilibrium – there has simply not been enough time since the Big Bang for larger systems of super-clusters to collapse and come to dynamical equilibrium. The largest galaxy clusters contain hundreds of galaxies that orbit at speeds of several thousand kilometres per second within a volume no larger than that occupied by the Milky Way and Andromeda. These environments are very hostile places for galaxies, which frequently suffer strong gravitational shocks as they pass rapidly by each other, stripping away stars and dark matter. Most of the mass of a galaxy cluster resides in a smooth dark matter component that has been gravitationally torn away from the individual galaxies.

Clusters are extremely useful systems for measuring the composition of the matter in the universe. They have formed from such a large region of space that they have collected a fair and representative sample of matter. (Astrophysical processes may have altered the baryonic content of smaller systems such as individual galaxies, for instance through the ejection of gas from supernova winds.) Moreover, there are several independent methods

**Figure 3.3.** The spectacular cluster of galaxies Abell 2218 deforms the fabric of space and acts as a giant gravitational lens. This image, taken with the Hubble Space Telescope, show many giant arcs of light that are in fact distant galaxies that have been distorted and magnified by the strong gravitational field of the foreground galaxy cluster. The positions and shapes of these images can be used to accurately reconstruct the amount and distribution of mass within the cluster.

for measuring their mass distributions: standard equilibrium estimators that use the galaxies as test particles; the hydrodynamics of the X-ray emitting gas; and gravitational lensing. In general, these methods give similar results and support Zwicky's original finding that clusters must be filled with large amounts of dark material.

For these reasons, clusters of galaxies are the systems of choice for quantifying the mean mass density of the universe and there are two ways in which to accomplish this. The standard method is to compare the ratio of 'mass to light' of clusters with the value required for a critical universe that comes to rest in an infinite length of time. Averaged over the entire universe, this critical density is equivalent to about one proton per cubic kilometre. (This method relies upon the assumption that galaxies are an unbiased tracer of the mass, i.e. the efficiency of galaxy formation does not depend upon environment.) The second method is to compare the 'mass to baryon' fraction inside clusters with the *theoretical* baryon density calculated from the thermal history of the early universe. (This method relies upon understanding nucleosynthesis during the Big Bang and the assumption that galaxy clusters contain a fair and representative sampling of material within the universe.) The latest results using either of these techniques suggest that the mean density of the universe is about a factor of three lower than that required to halt its inevitable infinite expansion.

The question that must be asked is: what is the largest scale on which dark matter is present? Is there a smooth component of mass that is clustered beyond the regions that can be probed by galaxy clusters? If this is the case then its properties or even its presence may be very difficult to detect. One possible candidate that comes in and out of favour with theoreticians is a component of 'hot dark matter' – particles that move so fast that they cannot cluster on small scales and may give rise to a smoothly distributed component of mass on large scales. One way to detect such a component is to take advantage of the fact that large mass inhomogeneities cause the motions of galaxies to deviate from their initial radial, or Hubble expansion. The so called 'peculiar motion' of galaxies can be measured and related to the underlying mass distribution in a well defined way. Unfortunately, this relationship is only simple if the galaxies are unbiased tracers of the mass distribution, which is probably not the case. Nevertheless, these analyses give similar numbers for the mean density of the universe as have been obtained using galaxy clusters and they provide comforting support that galactic motions can be explained purely by gravitational forces.

Current evidence is pointing towards a low value of the mass density, even when averaged over large scales. Higher mass densities would cause galaxies to have larger random motions than observed – both within clusters and on larger scales. Could the dark matter be responsible for suppressing galactic velocities? For example, if the dark matter interacts with itself via the strong interaction then the motions of galaxies would be suppressed by the cosmic drag or pressure that occurs as halos move through space. Yet again we turn to the prospects of using gravitational lensing to probe the distribution of mass. The weakly nonlinear structures in the universe distort the images of background galaxies in the same way as galaxy clusters or galactic halos. In this case, the distortions will be even weaker but the area that can be surveyed is considerably larger. Surveying a large enough region of the universe at the resolutions required to measure the tiny distortions in background galaxies requires a costly dedicated precision telescope. Nevertheless, such a survey would allow us to reconstruct the three-dimensional mass distribution and would be invaluable for testing cosmological models.

## Theoretical motivation

The detection of the cosmic microwave background (CMB) radiation firmly established the hot Big Bang model as the framework within which to construct detailed cosmogonic models. Observed today at $2.7277 \pm 0.0001$ kelvin (a remarkably precise measurement to find in astronomy), this is the current temperature of the relic radiation from the epoch of creation after 15 billion years of cosmological expansion has stretched and cooled the residual relic photons. The universe has evolved from a dense and extremely smooth 'hot fireball', into the complex distribution of stars and galaxies that we observe today. The tiny fluctuations that the Cosmic Background Explorer (COBE) satellite recently observed imprinted onto the microwave background reflect real fluctuations in the mass distribution at an early epoch. It is thought that galaxies have grown from fluctuations with a much smaller scale than these, imprinted in the cosmic mass distribution by some, as yet unknown, random quantum mechanical process.

The CMB radiation contains a wealth of information on the fundamental parameters that have shaped our universe. Two planned space-based experiments, the Microwave Anisotropy Probe (MAP) and the Planck Surveyor, will make precise temperature maps of the background radiation at much higher angular resolutions than COBE. Various physical mechanisms leave characteristic imprints on the CMB that can be used to measure combinations of the mean mass density, cosmological constant, baryon density and even the Hubble constant that determines the age of the universe. Recent experiments that have been carried on short balloon flights at the South Pole have already mapped a small region of the sky at high resolution and the preliminary reports are that the mean density of the universe has been accurately measured and is remarkably close to the critical value. By the year 2010 the CMB may have provided precision measurements of most of the fundamental parameters that govern the evolution of the universe, leaving astronomers with the task of fitting the pieces together to form the 'big picture'.

Evidence for the hot Big Bang is now overwhelming and 99.9 per cent of researchers believe that this model is basically correct. There are many fundamental problems that we do not yet understand such as: why particles have mass; why the universe ended up with more matter than antimatter; what caused the tiny fluctuations that seeded galaxy formation; why the initial expansion took place, etc. These problems are extremely complex yet we may see scientific answers to these questions within our lifetimes, rather than appealing to anthropic or religious arguments. The theory of 'inflation' predicts that we live in a flat universe that slowly grinds to a halt from its current rapid expansion. Inflation does not allow for a universe that re-collapses under its mutual gravitational attraction, thus resolving some of the need for a special cosmic creation. However, the total sum of everything we can observe, including the dark mass associated with galaxies and clusters, adds up to about one-third of the required mass. The cosmic inventory leads to the conclusion that the universe will expand forever, slowly fading away as the stars burn out and all of the mass becomes 'dark matter'.

The second ingredient that must be added to this 'standard' model is the dominant dark matter component. The nature and amount of dark matter govern the way in which structure forms within the expanding universe. Although some of the dark matter may be ordinary atomic material, or something far more exotic, the thermal history of the universe during the first few minutes of the relativistic expansion can be used to predict the total amount of baryonic (protons, neutrons) matter in the universe. Certain isotopes, such as deuterium, were only created during the Big Bang and their present day abundance can be used to measure the density of normal matter. The theory of 'primordial nucleosynthesis' predicts that the density of baryons in the universe should amount to just two or three per cent of the critical density, whereas the inventory of cosmic baryons has shown that galaxies contribute less than one per cent of the critical density. Thus, astronomers are faced with a second dark matter problem and a search for hidden baryonic matter.

## 3.2.1 Baryonic dark matter

Could some of, or even all, the dark matter be provided by regular baryonic material but locked into some form that does not emit light? Many candidates have been considered, from dark stars to frozen crystals of hydrogen. However, a host of observational and theoretical constraints have been used to eliminate practically every possible means of hiding a closure density of dark baryons. Depending on just how much dark matter in this form we are willing to accept, the most likely remaining possibilities are diffuse warm gas, cold and very dense clouds of molecular hydrogen or old stellar remnants of stars like the Sun.

Recent developments in X-ray astronomy have lead to the discovery that a large component of the dark mass within groups and clusters of galaxies is hot ionised hydrogen plasma. The first X-ray satellites that could map the distribution of ionised gas were sensitive only to the highest energy photons and temperatures beyond $10^7$ degrees – lower temperature gas was not detected until mapped by the ROSAT satellite. The data show that the ratio of mass to *light* within groups and clusters is typically as high as 300 times that of the Sun. However, once the diffuse gas is considered, then the mass to *baryon* ratio falls to about 30, close to the universal value predicted by nucleosynthesis.

It is difficult to envisage a scenario in which the missing baryonic dark matter associated with isolated galaxies is not in a gaseous form. It does not make sense that most of the baryons in galaxies are low-mass stars whilst the baryons in clusters remain as diffuse gas. Clusters form via the mergers of galaxies like the Milky Way, therefore the gas must already be present and subsequently shock heated to the equilibrium temperature in the denser environments. One possible location for this missing gas may be the high velocity gas clouds. Thought to be nearby gas clouds ejected from the galactic disc by supernova explosions, these clouds of neutral hydrogen may really lie at much larger distances and therefore be much more massive than previously thought.

If the universe contained just baryonic material, fluctuations as small as those observed in the CMB would not have had time to grow into objects as large as galaxies or galaxy clusters. The most popular idea to accelerate the growth of fluctuations is to appeal to a large density of matter in the form of 'exotic' particles created during the early universe. Whether we like it or not, both theory and data point towards an additional component of mass that is in some form other than the baryons from which we are made.

### 3.2.2 Exotic particle dark matter

Photons or neutrinos are examples of non-baryonic particles that we know exist. Non-baryonic dark matter candidates can be categorised according to their velocities when they form in the early universe and the forces by which they interact with other particles. 'Hot' dark matter is a good candidate because the neutrino is known both to exist and have mass. Unfortunately, its relativistic velocity is so high that the small fluctuations that could form galaxies are erased before they have time to collapse. Warm dark matter (perhaps heavy neutrinos) remains a viable candidate, but it is difficult to theoretically motivate a significant cosmological density of such particles. Cold dark matter (CDM) has proven to be the most popular candidate, stemming in part from its apparent lack of free parameters in its first conception, motivation from particle physics, and its success at reproducing a host of observational results.

The only convincing way of demonstrating that most of the material in the universe is made up of particles is to detect some of them in a laboratory. Two candidates remain as contenders for CDM, that are both promoted by theoretical particle physics models that attempt to unify the forces of nature. The axion has a mass similar to that of the electron, and the neutralino has a mass of about 100 protons. The expected density of dark matter at the Earth's position in the Galaxy, 30000 light years from the centre, is equivalent to one proton per cubic metre. Therefore at the Earth's velocity of $200 \, \mathrm{km \, s^{-1}}$ around the centre of the Galaxy, we expect about one thousand neutralinos passing through our body every second!

If neutralinos are to provide all of the dark matter, then they must have a tiny interaction cross-section, much smaller than the size of a hydrogen nucleus. Roughly once per day, a neutralino will collide with a nucleus in our body, raising our temperature by an immeasurable amount. Such a small energy release is only detectable in a super-cooled and highly pure material. Several research groups are using germanium crystals to detect temperature increases from individual collisions between nuclei and dark matter particles. Axions can also be detected in a laboratory by stimulating their conversion to photons using a very strong magnetic field. These laboratories are located deep underground, often at the bottom of mineshafts where conditions for carrying out precision measurements are not ideal, but the spurious signals from background events are lower than on the Earth's surface. One way to discriminate from noise and spurious events is to search for a yearly modulation in the signal that results from the motion of the Earth around the Sun. A highly controversial claim for such a signal has already been made by the Italian DAMA collaboration. If independent experiments with better statistics and longer baselines can verify this detection, then the implications will have a similar impact to that of the discovery of the microwave background.

Experiments searching for axions and neutralinos are underway which will probe the entire allowed parameter space within the next decade. However, current and planned direct detection experiments assume the dark matter in the galactic halo has a smooth distribution in the Galaxy. Numerical simulations of cold dark matter halos have demonstrated that this assumption is incorrect, revealing that the expected phase space structure of galactic halos is remarkably complex. The spectrum of fluctuations in the cold dark matter component allows smaller and smaller objects to collapse at earlier epochs. This may result in a hierarchy of clumps of dark matter down to extremely small scales, even below the mass of the Earth and possibly as small as grains of sand which would be almost impossible to detect on Earth.

### 3.2.3 Vacuum energy and cosmological constant

The cosmological constant in Einstein's equations acts like a component of dark matter that fills space uniformly. This component leaves a visible signature by causing an accelerated or decelerated expansion detectable as deviations in the redshift–magnitude diagram. If the distances to high-redshift objects are measured accurately enough then it is possible to measure the deceleration (or acceleration) of the universe. A positive cosmological constant would manifest itself as distant objects appearing to be accelerating away from each other. The most recent results using of order 100 distant supernovae, suggest that the effective energy density of the cosmological constant is of order twice that of the matter density – the combination of matter and vacuum energy is sufficient to close the universe!

The physical nature of this energy is still highly speculative, as is the reason why it should make a non-negligible contribution to the energy density of the universe that is similar to the matter density. One can always invoke anthropic arguments and the 'many universes' concept, that now and again produce a universe capable of creating and sustaining life. The cosmological constant was invented by Einstein to make a static model for the universe – at that time, the Milky Way was thought to be the only object in the universe. The cosmological constant has also been spuriously invoked to reconcile observational evidence that the universe was older than allowed from theoretical models, as well as explain the abundance of high redshift quasars. It may be a real component that corresponds to the energy of the virtual particles associated with the vacuum; however, we should learn from the history of its misuse and be careful before accepting a non-zero value.

## 3.3  Computational cosmology

One of the main goals of astronomy is to explain the origin and properties of galaxies, such as their kinematics or clustering properties on large scales. From the microwave background fluctuations we know that the universe was highly smooth, to one part in a million, at early times. However, there were tiny fluctuations present that may have been generated during the inflationary period and these have grown through a process called gravitational instability. The cosmological parameters and the identity of the dark matter determine the rate of growth of these over-densities. Unfortunately, simple linear mathematical theory cannot follow the highly non-linear processes that lead to the subsequent gravitational clustering and galaxy formation. It was not until the advent of powerful computers and efficient algorithms for simulating the formation of structure, that different cosmological models could easily be compared with observational data. Numerical simulations have played a key role in the interpretation of astronomical data since computers first became available for research in the 1960s.

One of the first applications of computers to astronomy was to test the collapse and stability of a cluster of galaxies modelled by several hundred point masses. Direct simulation is a powerful tool for following the motions of dynamical systems as they evolve under their mutual gravitational potential. The mass distribution is represented by a system of particles (representing dark matter, stars, galaxies etc.) that are continuously moved according to the gravitational force acting at each point. The exact solution scales as $N^2$, where $N$ is the number of particles, but techniques such as updating the forces from distant regions of space less frequently can reduce the computational scaling to order $(0)N \ln(N)$. Cosmological simulation began in earnest in the 1980s with the first investigations of structure formation in a universe dominated by hot dark matter. Starting with initial conditions that are set by a combination of observations and theory, the mass distribution is evolved by continuously calculating the growth of structure within an expanding coordinate system.

As the resolution or number of particles in a simulation increases, then we can resolve more complex regions with greater realism, although this requires using more integration time-steps that increase in proportion to the local dynamical time. If we weight the work done on the particles inversely with their natural time-steps, we find a potential gain of a factor of 50, one of the last algorithmic areas where an order of magnitude improvement is still possible. From this point on we must rely on the continual increase in performance provided by faster processors to gain more resolution. Computers double in speed and memory capacity every 18 months; however, a significant increase in computing power has recently resulted from adapting the $N$-body codes to run on parallel computers. At the current rate of algorithmic and hardware growth it is anticipated that by 2010 computers will equal the capacity and throughput of the human brain! Over the next few years we should see major breakthroughs in our understanding of the universe as a direct result of being able to simulate physical processes, such as star formation, at realistic resolutions.

Parallel computers have many hundreds of individual fast processors linked together with fast connections in an optimum topological configuration for maximising inter-processor communication speed. The largest available systems for scientific research have as many as 1000 processors and memory (RAM) capacities of 100 gigabytes, although some countries have access to much larger machines to play war games on. Theoretical research now requires similar resources as observational astronomy, with large dedicated telescopes serving the community. By the year 2010 we should achieve one of the ultimate simulation goals – to resolve the formation of individual galaxies within a cosmological volume, a task that requires parsec resolution in both the dark matter and baryonic components. Extrapolating wildly from the current rate of technology growth to the end of the twenty-first century, it should then be possible to correctly simulate the formation of stars and their individual planetary systems within a given cosmological context.

### 3.3.1 The hierarchical evolution of galaxies and clusters

Figure 3.4 shows the formation of a pair of cold dark matter halos that resembles the Local Group of galaxies. This simulation took many months of dedicated time on a large parallel computer and is the most intensive ever carried out for astronomical research (see http://www.nbody.net). More than $10^{15}$ operations were required to follow the evolution of the dark matter from the early universe to the present day. Initially there is no apparent structure in the mass distribution, although tiny fluctuations are present with amplitudes that are specified by the CMB observations.

The third snapshot in Figure 3.4 shows the universe at a redshift $z = 10$, when it was just 300 million years old. The mass distribution has a spectacular filamentary appearance, although a close-up view shows that the filaments are made up of many small dark matter halos. At this time the first stars and quasars would be lighting up and re-ionising the universe. Evidence for this activity may be found by searching for high redshift supernovae with the Next Generation Space Telescope. The amount of structure at this epoch is very sensitive to the nature of the dark matter. If the dark matter consists of 'warm' particles, then the planned space mission will detect nothing at these high redshifts. The first galaxy-sized halos are rarer and collapse at later times and can be seen forming at the intersection of the filamentary structures. The topology of the mass distribution is complex and in this particular model it is almost scale free – small structures appear as scaled versions of larger structures. At a redshift $z = 1$ the universe was approaching half of its present age (close to the epoch at

**Figure 3.4** (opposite). The hierarchical formation of a binary pair of galaxy halos in a universe dominated by cold dark matter. This plot shows a time sequence of six frames of a region of the universe that evolves into a structure that resembles our Local Group of galaxies. The brighter shades represent regions of higher dark matter density. Initially, the mass distribution is highly smooth but spectacular filamentary structures rapidly form. At the final epoch, ($z = 0$), the two large halos that would host the Galaxy and Andromeda are a million parsecs apart and moving towards each other at over $100 \, \mathrm{km \, s^{-1}}$. The simulation begins at a redshift $z = 50$ when the universe has expanded to just 2 per cent of its present size and 0.2 per cent of its present age.

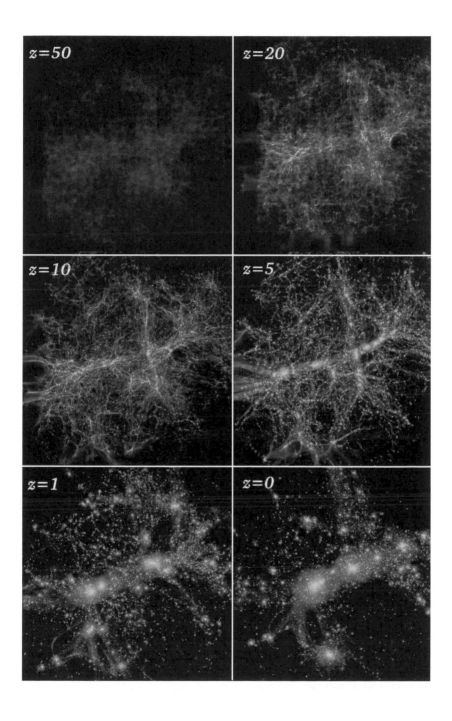

which the Earth formed from recycled stellar debris from supernovae), the system has almost completely formed. Although never fully in equilibrium since new mass continues to rain in, the galaxy halos undergo a further five billion years of internal dynamical evolution before they reach the present epoch at $z = 0$.

High-density regions such as galaxy clusters are the hardest regions to simulate because of the short dynamical timescales and strong gravitational forces. Only recently has it been possible to achieve sufficient numerical resolution to resolve the internal structure of dark halos. One of the most remarkable successes of hierarchical clustering models is the natural formation of clusters of galaxies. Starting from the spectrum of tiny mass fluctuations observed in the CMB, structures evolve by gravity alone to form massive dark matter potentials that contain hundreds of individual dark matter halos. The model also appears to be successful at reproducing systems that look like the Local Group of galaxies; however, upon closer inspection there are some serious flaws.

First, do discs that form within halos of cold dark matter have the same rotation curves as observed galaxies? This is a fundamental test of any dark matter candidate and unfortunately the CDM model appears to fail this test. In Figure 3.2 we plotted the rotation curve of a typical galaxy that is dominated by dark matter. The slowly rising rotation curve indicates a constant mass density in the central region whereas the density profiles of cold dark matter halos cause the predicted rotation curves to rise much more steeply than this. In other words, halos of cold dark matter have slightly too much mass in their central regions that would cause galaxies to be rotating much faster than expected.

A second problem with the CDM model that has recently been uncovered is the amount of substructure and small galaxies that are predicted to surround the Milky Way. Results from numerical simulations have demonstrated that cold dark matter halos are remarkably self-similar – small, low-mass halos appear as scaled versions of their massive counterparts. Although CDM clusters resemble real galaxy clusters, with many thousands of small halos embedded within a smooth background, galaxy halos of CDM appear exactly like galaxy clusters. Thus, the Milky Way should contain approximately one thousand satellite galaxies – not the mere dozen that are observed to orbit the Galaxy.

One solution to these problems is to appeal to some astrophysical process that can somehow suppress star formation within low-mass dark

**Figure 3.5.** A dark matter halo of cold dark matter particles that may surround the Milky Way galaxy. This simulation is the most intensive computer calculation ever performed for cosmological research, requiring over 150 000 cpu hours on a large parallel super-computer. The evolution of the dark matter proceeds in a nearly identical way as plotted in Figure 3.4, however, the mass resolution is much higher allowing us to accurately resolve the internal structure and substructure of a single dark matter halo. This plot is 600 000 parsecs across – a galactic disc would occupy the central 5–10 per cent of this structure.

matter halos. In this case, the substructure may be present within the Milky Way's halo, but be invisible to observations. The only way of detecting their presence would be through indirect kinematical tests such as the gravitational effect on the Galaxy's disc. In fact, the velocity dispersion of disc stars correlates with their age, with the older stars moving considerably faster than the stars that formed more recently. A physical heating mechanism for producing this correlation has been sought after without success. Dark matter substructure provides an efficient mechanism for heating stellar discs and could explain the observed age–velocity dispersion relation. The only problem is that the thousands of high velocity encounters between CDM substructure and the disc may provide too much disc heating.

A large part of the success of the CDM model stems from the fact that this model provides a picture of the universe that can successfully reproduce a lot of observational data. Its apparent failures on small scales that are just being uncovered indicate just how much effort has gone into testing this model against data. Many of its past problems have been overcome by the introduction of new concepts and additional free parameters that detract from the simplicity of the original model. Crucial to the success of CDM are the controversial concepts of *biasing* and *feedback*. The latter was initially invoked to resolve the conflict between the low observed peculiar velocities with simulation results, whilst supernovae feedback is necessary to darken galaxies and reproduce the flat luminosity functions observed in galaxy groups and the field. The structure and substructure of dark halos holds the key to the identity of the dark matter. Cold dark matter particles interact with each other only gravitationally and through the weak interaction – one interesting possibility would be to investigate structure formation in a universe dominated by strongly interacting particles. It is worth mentioning that the current observational data cannot distinguish between the extremes in which the dark matter is weakly or strongly interacting with itself.

## 3.4 Summary

The observational evidence supports the hot Big Bang model and a universe in which structure grows hierarchically, with small objects collapsing early that merge and accrete into the galaxies, clusters and super-clusters that we observe today. Currently, there is no viable alternative to this scenario. If observations discovered massive or old objects already in existence at high redshifts then we could question the hierarchical model. The growth of structure is governed by the nature of the dark matter and the fundamental cosmological parameters. A universe dominated by a critical density of cold dark matter was the first cosmological model that had real predictive power and could withstand scrutiny against observations. The standard version of this model, with its well motivated parameters provides a picture that is remarkably close to reality. Although there is some degree of failure on small scales, whether or not this is fatal for the cold dark matter model will only be established after further observational and theoretical investigation.

There are still some problems to resolve, such as the location of the missing baryonic material and the identity of the dark matter. The cosmic inventory reveals that the universe contains about one-third of the mass necessary to halt its rapid expansion. Recent observational data suggest that the vacuum energy density may actually be larger than the mass density. Although this confirms the standard inflationary theory of a closed universe, it raises the question as to why the cosmological constant has a measurable value – the natural expectation was that it was either zero or incredibly large. Is this added complexity just reflecting our ignorance of the nature of dark matter? Could an alternative dark matter candidate reconcile observations with our theoretical prejudice for a closed universe, without resorting to a cosmological constant?

Ultimately, gravitational lensing surveys, microwave background experiments and direct detection experiments will reveal the properties of dark matter and the parameters that are embodied within our cosmological world model. Although we search for a single candidate, nature may not be that kind and we may live in a universe that contains several species of dark matter. The recent discovery that the neutrino has mass implies that 'hot dark matter' contributes a mass density equivalent to all the visible stars in the universe! Particle physics and theory have motivated a wealth of additional candidates, from topological defects to ghost universes. Hopefully the identity of dark matter will be revealed within the next decade or so, by a combination of strategic observations and theoretical modelling.

## 3.5  Further reading

Moore, B. 1999 *Phil. Trans. R. Soc. Lond.* A **357**, 3259–3275.

# 4
# The hottest spots in space?

## Malcolm D. Gray

*Department of Physics, UMIST, P.O. Box 88, Manchester M60 1QD, UK*

## 4.1 Introduction

If we assume that an object is a perfect emitter of radiation, a blackbody, we can use its measured brightness, and our knowledge of the geometry of the observation, to recover an estimate of the object's temperature, called a brightness temperature. For many objects in astrophysics, stars and galaxies for example, the results are very sensible, but this is not always the case. One class of radio sources have brightness temperatures that can exceed $10^{12}$ K, around 100 000 times the core temperature of the Sun. Such remarkable temperatures become much more mysterious when we look at their origin. The radiation appears to be emitted by molecules which could not survive at temperatures much above 1000 K! How can this apparent contradiction be resolved, and are the brightness temperatures real? The key to understanding this paradox is to look for methods of generating very bright radiation which do not require the source to be very hot: we must throw away the initial blackbody assumption, and look in more detail at the internal structure of molecules.

Molecules in a gas typically have many internal energy levels due to various modes of motion, including vibration and rotation. Usually, we expect these levels to be populated such that comparison of the number of molecules, the population, in two of the levels, gives a lower population in the higher level. Under exceptional conditions, however, the reverse can be true, and a population inversion is then said to exist between the two levels. In the normal case, radiation interacting with a molecule is more likely to undergo absorption, and energy is lost promoting the molecule to the upper state. But in the inverted case it is more likely to produce stimulated emission, amplifying the radiation while the molecule descends to the lower state. The inverted case then provides the answer: a fairly cool assembly of molecules can amplify radiation, increasing its brightness to values which, on the blackbody assumption, would correspond to a much hotter source. The stimulated emission process gives the name maser (Microwave Amplification by Stimulated Emission of Radiation) to this class of sources. These natural masers have much in common with laboratory masers and lasers (lasers use visible light rather than microwave and radio frequencies) but lack the mirrored cavity which gives laboratory instruments their very precise beam and frequency characteristics.

It is only 35 years since the first astrophysical maser was detected. Initially, the source caused consternation and the substance responsible was dubbed 'mysterium'.Very quickly, the spectral line was identified as a transition of the OH (hydroxyl) molecule and an explanation of the signal provided in terms of the maser amplification process already described.

The main problem in identifying the original maser was that the molecular transition responsible is intrinsically very weak, and so not expected to be observable. One great advantage of maser sources is thus immediately apparent: the amplification process allows us to view small regions and weak transitions that we can observe by no other means. Astrophysical maser sources are so small that they usually have to be measured by an array of telescopes operated together to boost the angular resolution. Individual spots of emission can have angular sizes of 1mas (milliarcsecond) or less. If our eyes had the same accuity, it would be possible to look up at the Moon at night and pick out one of the old Apollo landing craft on the surface.

So far, maser emission has been detected from eleven different molecular species, and typically each molecule produces maser emission from more than one of its transitions, so the number of maser frequencies or spectral lines which have been detected outnumbers the list of molecules many times. The source regions of the masers include some of the most dynamic and enigmatic environments in astrophysics, including regions of star formation, nuclei of disturbed galaxies, and supernova remnants. Closer to home, masers are also found in comets and planetary atmospheres.

## 4.2 General characteristics

In the introduction, we have already looked at one method of identifying a likely maser candidate: a discrepancy between acceptable temperatures for the emitting molecules and the brightness temperature of the maser. Maser spectral lines have a finite width, related to the first of these temperatures, but become narrowed by the amplification process, which is most powerful near the centre of the line, leading to characteristically sharp lines.

When observing stars and planets, the spherical geometry of these objects is an enormous aid in the interpretation of the observations. By contrast, masers have no well defined geometry: interferometer maps reveal objects of irregular shape which makes them both intriguing and challenging. This problem is compounded by the fact that, unlike a star, the maser's radiation pattern could be very different from its physical geometry. A roughly spherical clump of gas, for example, could give rise to a highly beamed maser.

A population inversion is required in a maser source, and maintenance of the inversion requires a pumping mechanism. Three types of mechanism can contribute to the overall pump: collisions with partner atoms and molecules, in astrophysics usually a mixture of atomic and molecular hydrogen plus helium; radiation, usually of far infrared (FIR) wavelengths; and chemical reactions which, when forming the maser molecule, leave it in the upper state of a maser transition.

Maser sources are almost invariably small: not in an absolute sense, as many are Solar-System sized, some even light-years across, but they are very compact relative to the environment in which they are found. For example, in regions where stars much more massive than the Sun are forming, maser spots are smaller than the region by a factor of about 1000. The small relative size of the maser sources makes them extremely useful: they can give us diagnostics of their environments with unrivalled spatial precision.

## 4.3 Galactic masers in star-forming regions

Stars which are much more massive than the Sun also pour much more energy into their environment during their lifetimes, including their formation. As these stars form, some of their energy often goes to pump maser emission, which is observed to come from many transitions in at least six different molecular species, though not all from the same source. Masers from different species and different lines appear to inhabit rather different regions within a given source, and this reflects the different pumping requirements of the various transitions and molecules. Of course, as the young stellar object (YSO) at the heart of a source ages, conditions in its surroundings may evolve to favour a different set of masers.

Eventually the YSO becomes hot and begins to emit copious quantities of ultraviolet (UV) radiation which ionises the hydrogen in the surrounding gas, vaporises the icy mantles of dust grains and introduces a general outflow to the motion of the circumstellar material. OH masers form close to the boundary between the ionised gas and the surrounding molecular gas. The OH masers appear in several lines (see Figure 4.1) but the 1665 MHz transition is the brightest and most common. The sites of the OH

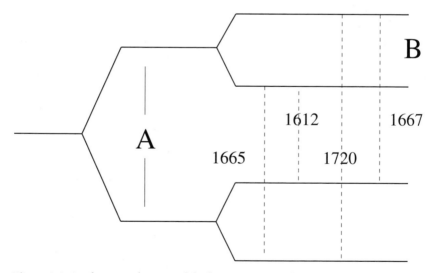

**Figure 4.1.** A schematic diagram of the lowest rotational state of the OH molecule. Transitions observed as masers have their frequencies marked in MHz. Energy splittings are due to: A an interaction between the rotation of the molecule and electronic angular momentum (Lambda doubling); B an interaction between the electronic spin angular momentum and the spin angular momentum of the hydrogen nucleus (hyperfine splitting). The splittings are not shown to scale. The next two higher rotational states are similar, giving rise to groups of potential maser transitions near 6000 and 4600 MHz respectively.

masers may be either in a disc of accreting matter around the YSO when the magnetic field is quite ordered, or in a more disturbed region. The siting probably depends on the evolutionary state of the YSO, which destroys the disc as its ionised region grows. Perhaps the most remarkable feature of OH masers in star-forming regions is that many of them are 100 per cent circularly polarised, though some elliptically and linearly polarised masers are also observed. Polarisation and the Zeeman splitting of their spectral lines allows us to learn a great deal about the magnetic fields local to the masers. In particular, amplification along the magnetic field lines is the probable means of developing very high degrees of circular polarisation. Sources which are probably disc-like can be analysed using the polarisation properties of maser spots. The apparently disordered field of the W75N source can be understood on the basis that we see masers from both the topside and underside of a circumstellar disc.

Methanol masers in star-forming regions can be divided into two distinct types, class I and II, depending on the set of transitions present. The class II sources, like OH masers, appear to be close to ionised regions. In some cases, OH and methanol regions appear to be close together and may even overlap. Groupings of class II masers are linear, with a velocity gradient consistent with rotation and these objects are probably associated with accretion discs around massive YSOs, like the younger OH sources. Variability is on similar timescales to that of OH, with changes typically over years. Class I sources by contrast appear spectrally and spatially consistent with turbulent motions and may mark an even earlier stage in the evolution of the YSO, prior to significant ionisation.

Some of the energy injected into the interstellar medium (ISM) by YSOs is in the form of sound waves, which break and form shock waves in the lower density surroundings of the star. If these shock waves are not severe enough to dissociate water, they generate conditions in the gas behind the shock wave which are ideal for forming water masers via a predominantly collisional pumping scheme. Water masers are invariably associated with shocks in star-forming regions. Like OH and methanol, water can mase in many transitions, ranging in frequency from 22 GHz up to at least 658 GHz. The shock waves which form the water masers can be consistent with generation by supersonic turbulence in some sources, but in others they appear closely linked with the positions where bipolar outflows from the poles of the YSO interact with the background medium. Water masers in star-forming regions have the highest brightness temperatures of all types and the luminosities can exceed the entire output of the Sun. They are also highly variable on timescales of weeks to months. Sometimes a burst occurs, in which a maser feature can become brighter by a factor of ten or more.

## 4.4 Galactic masers in circumstellar envelopes

When stars reach an advanced stage of evolution, they become cooler and swell to form objects known as red giants or red supergiants, depending on their mass. These stars eventually reach a stage where they become variable and, at least for the lower mass stars, the variability is pulsational. The pulsations drive shock waves into the stellar atmospheres, which become dense and undergo rapid mass loss, which can exceed one Earth mass per year. These dense, distended atmospheres or circumstellar envelopes (CSE) are the sites of a rich assortment of circumstellar masers. As the stars evolve further, the envelope structure tends to become more opaque and changes in composition. One class of star, the OH/IR stars, are identified through their maser and infrared emission; their envelopes are completely opaque to visible light.

The maser species present in a CSE depend strongly on its composition, which is a function of the evolutionary state of the star. Initially, envelopes have an excess of oxygen over carbon (M-type or oxygen-rich): chemical reaction networks dictate that all the carbon atoms in the CSE are bound as CO and the masers observed are based on the oxygen-based species SiO, water and OH. As the star evolves, internal convection raises recently nucleosynthesised carbon to the surface and, after passing through an intermediate (S-type) phase where the carbon and oxygen abundances are similar, the CSE becomes carbon rich (C-type). C-type CSEs have hydrogen cyanide (HCN) and, very rarely, CO masers. Throughout this sequence, stellar evolution proceeds ever more rapidly and consequently we see far fewer C-type envelopes than M-type.

Stars with M-type envelopes have up to three concentric zones of maser emission. Images taken with global interferometer networks have shown that the SiO masers lie closest to the stellar surface, sampling the densest, hottest and most energetic region of the envelope, within about three stellar radii of the surface. Further out, at roughly ten stellar radii from the surface, are water masers, detected at several frequencies, and closely associated OH masers at 1665 and 1667 MHz. Finally, at 50 or more stellar radii, lies the outermost shell, composed of 1612 MHz OH masers.

The development of the three shells is probably also evolutionary, developing from inside to outside, and certainly many stars have only one or two of the shells.

Interferometer measurements of the 1612 MHz masers with instruments like MERLIN show that the structure is that of a thin spherical shell, dominated by radial amplification: the caps of the shell in front of, and usually behind, the star appear much brighter than the limbs. The double-peaked spectral structure of these masers indicates that they are situated in the outer, smoothly outflowing region of the CSE. A useful method of estimating distances to these stars, based on the phase lag between variations of the starlight and maser, has been developed (Figure 4.2).

Proper motions of water masers are studied in an intermediate zone where outflow has been established in the CSE, but where it is still subject to shocks and probably to turbulent motions. There are many water frequencies emitted, some of which are probably very closely associated spatially, and others which are not. Interferometer observations show the masers grouped in a rather indistinct ring structure, indicating a combination of radial and tangential amplification. The OH mainline masers

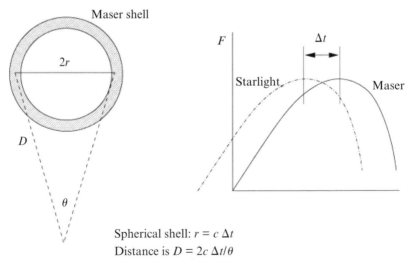

Spherical shell: $r = c\,\Delta t$
Distance is $D = 2c\,\Delta t/\theta$

**Figure 4.2.** A method of measuring stellar distance using 1612 MHz maser shells. An interferometer, such as MERLIN, measured the angular diameter, $\theta$, related to the true diameter, $2r$, and the distance, $D$. The radius, $r$, is calculated based on the time-lag between changes in the starlight and the maser, based on the light travel time from star to shell.

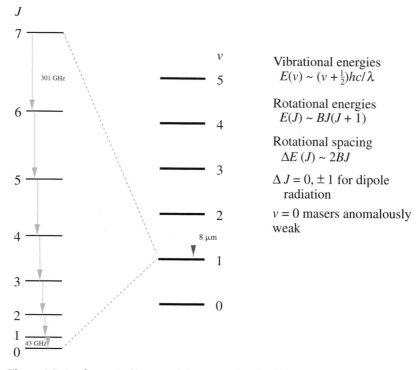

**Figure 4.3.** A schematic diagram of the energy levels of the SiO molecule. Roughly equally spaced vibrational states are linked by radiation of wavelengths near 8 μm. Within each vibrational state are a stack of rotational states with increasing quantum number $J$. $B$ is the rotational constant of the molecule. Masers have been observed from the vibrational states 0–4, and those transitions observed as masers in the $v=1$ state are marked. Frequencies range from 43 GHz for $J=1$–0 to 301 GHz for $J=7$–6.

probably form from the dissociation of water molecules by ultraviolet starlight at the outer edge of the zone. The proper motions of these masers, combined with a suitable model of the geometry can yield stellar distances via a method called moving cluster parallax.

The structure of the energy levels of SiO, with masing transitions marked, is shown and explained in Figure 4.3. Masers have been detected from vibrational states $v=0$ to $v=4$, but those in $v=1$ and $v=2$ are the strongest. An example of a 300 GHz $J=7$–6 maser spectrum is shown in Figure 4.4. Interferometer images show that, at least at 43 GHz ($J=1$–0), the masers form a ring structure around the star which varies with time, with

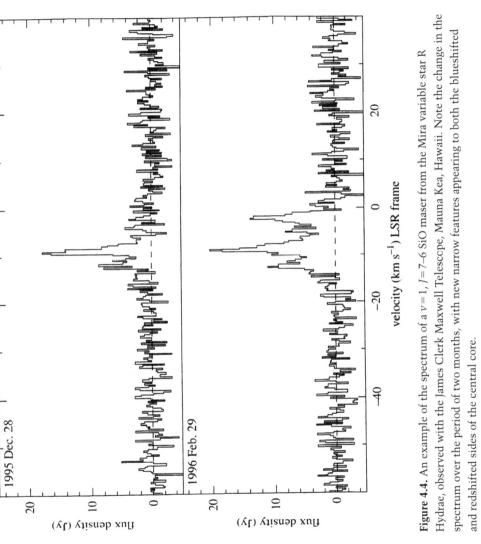

**Figure 4.4.** An example of the spectrum of a $v = 1$, $J = 7–6$ SiO maser from the Mira variable star R Hydrae, observed with the James Clerk Maxwell Telescope, Mauna Kea, Hawaii. Note the change in the spectrum over the period of two months, with new narrow features appearing to both the blueshifted and redshifted sides of the central core.

spots directly in front of the star being rare and weak. This suggests that amplification is predominantly tangential. In the SiO maser zone, the outflow and mass-loss processes are only just being established: the dominant motion of the envelope is pulsational, and both expansion and contraction of 43 GHz maser rings have been detected. The whole structure of the SiO zone appears to be intimately tied to the periodicity of the star. Hydrodynamic models of the envelope, supported by observations, suggest that shock waves sweep outward through the envelope, so that a clump of gas supporting an SiO maser should experience a sequence of violent outward accelerations by the shock waves. A computer simulation of this shell has also been produced with the ultimate aim of calculating mass-loss rates as a function of radius and time. Some frames from this film are shown in Figure 4.5(a)–(d). Readers with access to the Internet can view the whole animation at http://www.astro.cf.ac.uk/pub/Malcolm.Gray/sio/siomed.gif with a counterpart for a real star (TX Cam) at http://www.jb.man.ac.uk/~pdiamond/txcam2.html.

Polarisation is frequently observed in SiO masers and is usually linear. A very interesting experiment would be to follow the planes of polarisation through a stellar cycle, since a rapid change of the polarisation plane would probably fix the position of the shock wave at any phase of the stellar cycle. Tracing the shock wave position is vital for fixing the phase relationship, at present poorly known, between the optical light curve (phase zero at maximum light) and theoretical models (phase zero at maximum outward velocity of the sub-photospheric layers driving the pulsation). One possible method is to use high frequency SiO masers, for example $v = 1$, $J = 7$–6 at 300 GHz (Figure 4.4), which require very high temperatures only found just behind the shock wave, to trace its position. However, at present no telescope exists which is capable of producing high-resolution interferometer images at such a high frequency.

Hydrogen cyanide is the only strong maser found in C-type CSEs. There appear to be several frequencies, but the orginal detection, in a vibrationally excited form of the molecule, at 89 GHz appears to be present in envelopes with moderate mass-loss rates. Detection rates in a survey were about 20 per cent.

Another maser in the vibrational ground-state has also been detected towards C-type stars which are optical stars – they have quite thin envelopes with very low mass-loss rates. There is therefore a problem in observing the heavily obscured carbon stars, which at present have no maser tracer. The frequencies of the HCN masers are unfortunately just a little too high to be traced with present very long baseline interferometry (VLBI) networks in the same manner as SiO, so the full diagnostic power of these objects must also wait for an improvement in interferometer technology.

Although CO is a very abundant molecule in CSEs, as well as the rest of the ISM, there is, to date, just the unique maser source in the envelope of V Hydrae. The pumping mechanism has been analysed, and the most important factor is the extremely high CO abundance: the opacity in two vital FIR pumping lines is sufficient to destroy the pumping mechanism by weighting the decay of a high lying energy level away from the upper level of a potential maser, severely weakening the pump.

## 4.5 Megamasers

Masers of some of the types mentioned above have also been detected in nearby galaxies. Megamasers, however, are a completely different source type, which is far more energetic than the Galactic masers discussed so far. The host galaxies of megamasers all appear to be peculiar or disturbed in some way. They may be tidally perturbed, undergoing an intense burst of star formation or a merger event, or have a non-thermal active galactic nucleus (AGN). Galaxies with AGN are believed to be powered by a super-massive black hole in the core. The common factor between all these apparently disparate host galaxy types is that they are all ultraluminous in the infrared which allows for efficient searches. This emission is almost certainly from dust which absorbs ultraviolet radiation from the active region and re-radiates it in the infrared. Protected from the ultraviolet by the dust are molecules of water and OH which form the megamasers.

Apart from the obvious difference in overall luminosity, OH megamasers differ in several important respects from Galactic masers. The spectra are very broad, typically several hundred $\mathrm{km\,s^{-1}}$ wide. Sometimes there are additional wings, blue- and redshifted from the central emission core, separated from it by several hundreds more $\mathrm{km\,s^{-1}}$, so that the width of the entire spectrum can exceed $1000\,\mathrm{km\,s^{-1}}$. The dominant OH line in megamasers is the 1667 MHz line, rather than the 1665 MHz line in Galactic masers. The OH megamasers are less polarised and no Zeeman splittings have been measured.

Interferometer measurements show that the clouds which contribute to the megamaser spectra are only a few light years across, and in some

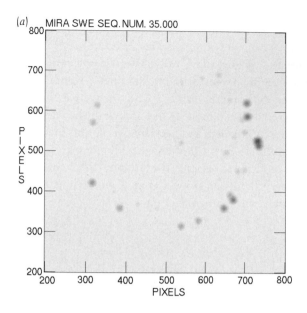

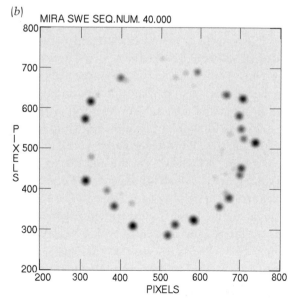

**Figure 4.5.** A sequence of computer simulated images of SiO masers at 43 GHz $(v = 1, J = 1–0)$ in the pulsating atmosphere of a Mira variable. Each pixel is about 1/20 of an Astronomical Unit. The period of pulsation of the modelled star is 1 year and the sequence of images covers 1/3 of the period, at intervals of

(c)

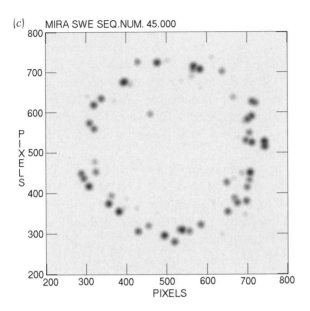

(d)

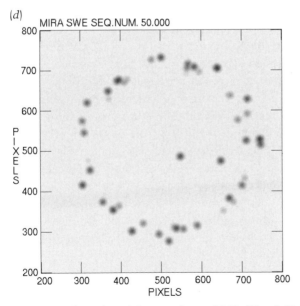

approximately 30 days. (a) is at a phase of 0.58; (b) at 0.67; (c) at 0.75; and (d) at phase 0.83. The phases are given as fractions of the period. Note that the modelled masers both move and change in brightness over the timespan of the model.

objects can be seen to have well ordered motion. In the case of water megamasers, the motion traced can often be understood in terms of rotation as
part of a disc around the central black hole. Binding masses can be determined from the rotational speeds, distances from proper motions and magnetic fields appear to be weak. In the spectacular case of NGC4258, the
binding mass for the central object was $3.5 \times 10^7$ solar masses. In the case
of NGC3079, the motion is much less ordered and has been modelled as
rotation plus strong turbulence.

OH megamasers also sometimes appear to be formed from elements of
a rotating molecular torus. Recent observations of MKN273 with the
MERLIN interferometer allowed the calculation of a central density for the
nucleus of some 10000 times greater than in the solar neighbourhood on
the basis of a rotation curve (see Figure 4.6).

## 4.6 Supernova remnants

A relatively new source type is the interaction region between supernova
remnants (SNR) and dense molecular gas. In a recent survey, no fewer than
33 detections of 1720 MHz OH masers were made from a sample of 75
SNRs. When correlated with observations of CO-rich gas near the maser
sites, the masers appear to be associated with thin molecular filaments
that correlate in position with synchrotron continuum emission, generated by relativistic electrons gyrating in a magnetic field. Models have
shown that the masers have a mixture of chemical and collisional
pumping. The idea is that magnetically modified shock waves form water
behind them, which is then photodissociated by soft X-rays from the SNR
to yield the necessary OH. The SNR masers had many spots at the resolution of the VLA (Very Large Array) interferometer. The VLA observations
had full polarization information and revealed quite a weak magnetic field
of $20 \mu$T. The velocity spread of the spots was a few $km\,s^{-1}$ and there is a
suggestion that the maximum amplification occurs when the acceleration
due to the SNR shock wave is perpendicular to the propagation direction
of the maser.

## 4.7 Other sources

If the sources discussed so far all seem very remote, there are astrophysical masers much closer to home, within our own Solar System. The recent

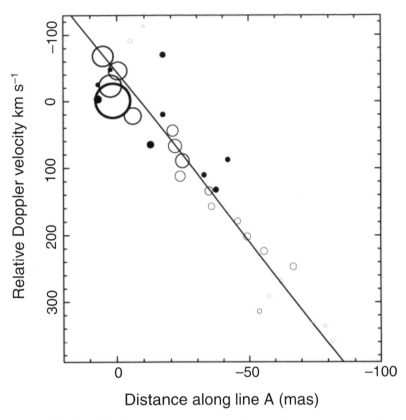

**Figure 4.6.** A plot of the OH maser motions in the core of the megamaser galaxy
MKN273. Radii of spots or circles are proportional to the brightness of the
features. Note that measured velocities of maser clumps follow a close line,
indicating mainly ordered motion. The velocities are also proportional to distance
along a line, A, estimated to be parallel to an edge-on disc. The motions with
velocity directly proportional to distance are consistent with a solid-body type of
rotation, expected for the mass distribution in a galactic nucleus. This plot can be
used to estimate the nuclear density and binding mass. The binding mass
corresponds closely to that of the supposed supermassive black hole in MKN273.

impressive show provided by the comet Hale–Bopp in 1997 had its radio
counterpart in OH maser emission at 1667 MHz, observed between late
February and the beginning of June. Interestingly, after the main cometary
emission had died away, a new spike appeared, where the cometary coma
had amplified a background source (see future observations, Section 4.9).

The final type of astrophysical maser needs no molecules, and hence no inversion in the usual sense of the word. These are free electron masers, and most are based on the cyclotron mechanism. An electron moving perpendicular to a magnetic field gyrates around the field lines with a characteristic frequency, the gyro-frequency, which is proportional to the magnetic field strength. An electromagnetic field which has a frequency resonant with the gyro-frequency can amplify at the expense of the energy of the electrons, so cyclotron masers appear at integer multiples of the gyro-frequency, but amplify best for small multiples. Since no molecules are involved, free electron masers can appear from very hot source regions, for example solar flares and, more generally, from the magnetospheres of stars and planets. Using the highly useful ability of maser amplification to make very small objects look very bright, a search has been made for extra-Solar System planets on the basis of their cyclotron maser emission, and several large-planet or brown-dwarf star candidates have been detected.

## 4.8 The theory of astrophysical masers

The gases in which astrophysical masers form are non-equilibrium systems, which means we cannot use the Boltzmann formula to calculate the fraction of the total population of the molecule of interest which will be in each of the energy levels of the molecule; if we could, there would be no inversions, and no masers! With the easy option removed, the level populations must be calculated by using master equations. These keep account of all processes which transfer population into and out of each level, to and from all the others. If the overall inward and outward rates are equal, we have a steady-state problem, if they are not, we have a time-dependent problem. The processes which move populations are a combination of collisions, non-maser radiation processes and the effects of maser lines themselves (saturation). Obviously, the more levels we wish to include in the model, the more complex the problem; analytic solutions are rarely possible for anything other than an idealised two-level system in a simple imposed geometry.

Models worthy of note should go beyond these assumptions. Efforts to include the magnetic-field dependence and polarisation in OH, so obviously necessary from observations, have been fairly successful. It is also desirable to have models where the radiation transfer is solved exactly, rather than making the usual local approximation. Such models have been applied to OH and to $H_2O$. Real maser sites are invariably dynamic environments, so velocity fields are another important ingredient, leading to new effects such as bifurcation of lineshapes when saturation is catastrophically lifted, and secondary gain, in which amplification down a velocity gradient is higher than through stationary material because of the additional molecules drawn into interaction with the radiation by the Doppler effect. Of course a solution for our chosen set of energy level populations is only as good as the molecular data which goes into the model. Obtaining these data requires interdisciplinary collaboration between astronomers and theoretical chemists. Collision cross-sections for the maser molecules with H and $H_2$ are hard to come by and are difficult to calculate. In spite of all the uncertainties, modellers have been remarkably successful in deriving pumping schemes for most of the maser species. It is possible in principle to trace the pumping scheme via recording the most significant computational operations made during the final iteration of a solution, allowing more exact characterisation of pumping mechanisms in the future.

## 4.9  The future of theory and observations

In terms of theory, advances should, and probably will, come in the direction of making models more realistic: old approximate methods will increasingly become replaced by models which are fully self-consistent, include chemical networks to produce the maser molecules, are time dependent, and have much less restriction to well defined geometries such as spheres and cylinders. For some special problems, maser initiation for example, fully quantum-mechanical treatments, where both the radiation and molecular ensemble are quantised, are desirable.

Increasing computer power will enable models to become ever more sophisticated. Indeed, without this numerical improvement, much of the progress discussed in the first paragraph would not be possible. However, the computer models should also become more clever, for example by automatic tracking of important elements in the pumping scheme, so that a summary of why the solution looks the way it does is readily available to the investigator.

One area in which I am extremely interested is generating a theory of broad-band maser radiation which is sufficiently accurate to address the problem of a proper coherence function for astrophysical masers. Such a theory would be complementary to a difficult but feasible experiment which could actually prove that astrophysical masers arise via stimulated emission. What is required is a resolution that picks out the response of individual velocity subgroups of molecules; this is about 0.01 Hz for 1665 MHz OH masers. Observing over such a tiny bandwidth should show whether or not the radiation has the properties of chaotic or coherent light. In fact, the coherence should be partial, and predictable by the accurate theory. Any result which fitted the partially coherent form, rather than the chaotic form, would prove that very high brightness temperature sources truly are masers.

The first observational development which will revolutionise maser astrophysics is a steady improvement in interferometer technology. Already, networks like the American VLBA and European EVN systems are capable of semi-automatic operation, allowing time-series observations of objects like the SiO masers in CSEs. Again, computing power is vital: quantities of data are now being routinely processed which would have been unthinkable just a decade ago. Observations are now also routinely phase referenced (a nearby continuum source is used to calibrate the phase of the masers) which allows very high absolute position accuracy. This is extremely important in that, for the first time, it allows us to ask realistically the question of whether two masers at different frequencies really amplify through the same column of gas and can therefore be used to test theories of competitive gain, where two or more transitions compete for the available inverted population. Resolution is also likely to improve as more antennas are based in orbit.

Interferometry will make another huge advance with the advent of the Atacama large millimetre wave array (ALMA); a huge international instrument with baselines up to 10 km which will be able to operate at up to the highest submillimetre frequencies available to ground-based observers. This object will vastly improve the positional information for all astrophysical masers above $c.$ 80 GHz.

The recent flight of the ISO satellite allowed observers to look not at the maser lines of OH, but at the FIR lines which pump them. Where results have been compared to the predictions based on pumping schemes required to sustain masers, agreement has been encouragingly good. There is also a possibility that FIRASERs, or far-infrared lasers, may exist well beyond the highest frequencies observed from the ground. Current observations have been inconclusive, but a successor to ISO is planned!

Megamasers are extremely bright and can therefore been seen at great distances, or equivalently, at high redshifts, so they are important cosmologically. There are two important tests on the evolution of the universe which can be applied using megamasers. The first is to study the rate of merging events in young galaxies. Remember that mergers are an important sub-class of megamasers, so the number of detected mergers from megamaser observations can be checked against the predictions of cosmological theories. The second test is to use megamasers to study molecular abundance in the early universe. The oxygen in OH and water must come from nuclear processing inside stars, so it would be expected to be rarer in young galaxies, so affecting the abundance and/or intensities of megamasers at high redshift. The problem with such observations is that the redshift moves the target frequency range down to the 200–1000 MHz range, where pollution from commercial communications equipment is already making ground-based radio astronomy very difficult. What is required is a fully space-based radio observatory.

A space-based system is definitely required to observe in the last great unexplored observing band. The ionosphere which allows us to bounce short-wave radio signals around the world also bounces back radiation at similar frequencies from space, so we can see nothing at frequencies below about 10 MHz. We do not expect to see molecular emission at these very low frequencies (VLF), but cyclotron and other free electron masers could be very bright. Recall that we expect masers at multiples of the gyrofrequency, which is proportional to magnetic field strength. Earth-like planets have fields which might produce cyclotron masers in the VLF range, so the VLF observatory would be an important step in the search for extraterrestrial life. An instrument to be based on the far side of the Moon has been mooted by the European Space Agency (ESA).

Astrophysical masers are natural amplifiers. Would we be the first civilisation to think of using them to broadcast our presence to others? With only limited power available, an effective way of boosting one's signal to the rest of the Galaxy would be to amplify it through a natural maser region. It could be, for example, a known water maser. The much geometrically diluted signal would then get a free amplification by a factor of up to $10^{12}$ times before going on its way. Remember that a natural background signal was detected after being amplified through the coma of comet Hale–Bopp. Perhaps some future element of the SETI programme should look at star-forming regions, not because of what is in them, but because of what might be behind them.

## 4.10 Further reading

Cohen, R. J. 1989 Compact maser sources. *Rep. Prog. Phys.* **52**, 881–943.

Elitzur, M. 1992 *Astronomical masers*. Dordrecht: Kluwer.

Gray, M. D. 1999 Astrophysical masers. *Phil. Trans. R. Soc. Lond.* A **357**, 3277–3298.

# 5
# Our Solar System and beyond in the new millennium

Andrew J. Coates

*Mullard Space Science Laboratory, University College London, Holmbury St Mary, Dorking RH5 6NT, UK (ajc@mssl.ucl.ac.uk, www.mssl.ucl.ac.uk)*

## 5.1 Introduction

In the seventeenth century, when *Philosophical Transactions of the Royal Society* was founded, the telescope had only just been invented. Humankind knew of six planets including our own. The next three centuries added Uranus, Neptune and Pluto to the known list as well as the many moons, asteroids and comets that we know today. Discoveries such as that Earth was not the centre of the universe and that planets orbit the Sun were key steps in increasing the understanding of our place in space. But it was only in the latter part of the twentieth century that we were privileged to carry out *in situ* exploration of the planets, comets and the solar wind's realm and to begin to understand the special conditions on Earth which meant that life started here.

In this article we briefly review our present knowledge of the Solar System we inhabit. The Sun itself, the planets, comets and asteroids are all discussed in general terms, together with the important discoveries from space missions which have led to our current views. For each of the bodies and for the interplanetary medium we present our understanding of the physical properties and interrelationships, with questions for further study.

We describe the solar wind and the way that it interacts with the planets and comets that it encounters. The importance of the obstacle in the magnetised plasma flow is particularly explored. We identify the gaps in our knowledge in each case.

What is in store for planetary exploration and discoveries in the new millennium? Already a sequence of Mars exploration missions including sample return, a landing on a comet, further exploration of Saturn and the Jovian system and the first flyby of Pluto are planned. We examine the major scientific questions to be answered and speculate on possible space exploration in the future. We also discuss the prospects for finding Earth-like planets and life beyond our own Solar System.

## 5.2 The Solar System in the last four millennia

Astronomy is one of the oldest observational sciences, existing for about four millennia. This time is estimated on the basis that names were given to those northern constellations of stars which were visible to early civilisations. Two millennia ago, a difference between the more mobile planets and comets on the one hand and the fixed stars on the other was realised by Ptolemy but not understood. At the dawn of the last millennium, six planets (Mercury, Venus, Earth, Mars, Jupiter and Saturn) were known. Aurorae were seen on Earth and recorded but understanding still eluded us. In the first decades of the last millennium comets were seen as portents of disaster. Little further progress was made during the Middle Ages.

Science began to take giant leaps forward in the sixteenth century. Copernicus realised that Earth was not at the centre of the Solar System. Tycho Brahe's planetary observations enabled Kepler to formulate laws of planetary motion in the seventeenth century. In the same century, Galileo

**Figure 5.1.** Our view of comets at the beginning and end of the last millennium is illustrated by a scene from the Bayeaux tapestry (left) and an image from the camera on ESA's Giotto spacecraft in which the Sun is towards the left (Giotto image courtesy of MPAe/ESA).

invented the telescope and Newton developed his theory of gravitation. Seven years after *Philosophical Transactions* was founded Newton demonstrated his interest in technology by inventing his reflecting telescope. The eighteenth and nineteenth centuries saw increasing use of this new technology, which deepened and increased understanding of the objects in the sky. Amongst many other discoveries were the periodicity of Halley's comet and the existence of the planets Uranus, Neptune and Pluto – one planet per century including the last one. The observations were all made in visible light to which our eyes are sensitive and which is transmitted through our atmosphere. The nineteenth century also gave us the basics of electromagnetism.

In the twentieth century we observed many important scientific and technological advances. In terms of astronomy we gained the second major tool for the exploration of the universe in addition to the telescope – namely the spacecraft. Space probes have not only opened up the narrow Earth-bound electromagnetic window which only allows us to detect visible light and some radio waves from the ground, but they have also allowed *in situ* exploration and sampling of our neighbouring bodies in the Solar System. Using the techniques of remote sensing to look back at the Earth has added a new perspective. For the first time we can now begin to understand our place in the universe and the detailed processes of the formation of the universe, our Solar System and ourselves. We are truly privileged to be able to use these techniques to further this scientific understanding.

We now know that, far from being empty, interplanetary space is filled with a hot, fast-flowing plasma, the solar wind. A plasma, sometimes called the fourth state of matter, is a collection of electrically charged particles which behaves on the large scale as a magnetised fluid while on the small scale the motion of individual charged particles is important. The Sun and other stars are so hot that their material is in the plasma state; in fact, over 99 per cent of the volume of the universe is plasma. Studying the solar wind and how it interacts with Solar System obstacles *in situ* allows us to study this important state of matter without the walls which confine and dominate Earth-based plasmas. As a result we can begin to understand how aurorae and comet tails form and to study the potentially dangerous effects of solar storms on humankind.

At the start of another millennium, the sense of wonder in looking at the night sky has not changed. As we use better ground- and space-based telescopic techniques and more detailed *in situ* exploration one can only feel a sense of excitement at the discoveries that the new millennium may bring. In this paper we will review our Solar System in general terms and, in particular, consider the interaction of the solar wind with the various planets and comets which inhabit the Solar System. We consider scientific questions for the future and speculate a little on how they will be answered.

## 5.3 Building the Solar System

The objects in the Solar System now reflect the history of its formation over 4.5 billion years ago. Because the planets are confined to a plane, the ecliptic, it is thought that the Sun and planets condensed from a spinning primordial nebula. The heavier elements in the nebula are thought to be present due to earlier nearby supernova explosions. As the nebula collapsed and heated, the abundant hydrogen fuel ignited in the early Sun. Gas further from the centre of the nebula became progressively cooler and condensation occurred onto dust grains. This caused differences in composition due to the progressively cooler temperatures away from the Sun. Gravitational instabilities then caused the formation of small, solid planetesimals, the planetary building blocks. Accretion of these bodies due to collisions then formed the objects familiar to us today.

This model predicts different compositions at different distances from the Sun, and this is seen in the different classes of Solar System objects today. The outer planets are associated with their much colder planetesimals, some of which remain as comets. The inner, rocky planets are associated with their own planetesimals, of which the asteroids are the partially processed survivors in the present Solar System.

For each of the present day Solar System objects we now briefly review the information which has been found from space missions and consider the major outstanding questions for future exploration.

## 5.4 The Sun

The Sun is an average small ('dwarf') star. In terms of its position half-way along a galactic spiral arm and its temperature (and consequently its luminosity) it is quite unremarkable. The Solar System with which it evolved is in our view remarkable, particularly since conditions were right for life to begin on the third planet some four billion years ago and to evolve from there.

The Sun is remarkable to us as it has a large enough disc for us to observe and the radiation we receive is intense compared to that from distant stars. Consequently, by studying the Sun we can try to understand how average stars work.

The present model of the solar interior is that hydrogen fusion reactions burn in a hot (15 million K), dense (over 100000 kg m$^{-3}$, or several times the density of lead), gaseous core. The core is depleted in hydrogen abundance compared to the remainder of the interior due to fusion reactions. Outwards from the core the density and temperature decrease rapidly. Heat is radiated outwards for up to about 80 per cent of the solar radius until the convection zone begins. The visible surface or 'photosphere' has a temperature of 6000 K.

The photosphere, and above it the chromosphere and corona, have energetic large- and small-scale structures organised by the magnetic field. Large-scale structures include the coronal holes, whose extent wanes with the increasing solar cycle, and which are the source of the high-speed solar wind which escapes along the magnetic field. Smaller-scale structures include coronal loops, sunspots, prominences and filaments. Short-lived features causing flares and coronal mass ejections are also important in determining the electromagnetic environment.

By comparison with other, similar stars we know that when the Sun runs out of hydrogen fuel in about five billion years the core will start to support other nuclear reactions involving carbon, nitrogen and oxygen. The core will contract but the envelope will expand to beyond the Earth's orbit. At that point the outer Solar System will be warmer than its current temperature and the situation may prevail long enough for life to develop on Europa or Titan.

## Our star, the Sun

Much of our present picture of the Sun, including models of its structure, has emerged from space missions. Most of the energy of sunlight, 1370 W m$^{-2}$ at Earth's orbit, and the spectral information it contains is only visible to space-borne detectors. However, there are several areas for which there are significant outstanding questions, as follows. How are transient events such as coronal mass ejections triggered? Will we ever be able to forecast their onset? Why is the temperature of the corona, at a million kelvin, so much hotter than the visible surface? Are our models of the internal structure correct? Why are there fewer neutrinos from the burning core than expected? How is the solar magnetic field produced? We await the space missions of the new millennium to answer these questions and inevitably to pose more.

## 5.5 The inner planets

Due to the temperature variation in the collapsing primordial nebula, the inner regions of the Solar System contain rocky planets (Mercury, Venus, Earth and Mars). The condensation temperatures of the minerals forming these planets were higher than the icy material in the outer Solar System. While it is fair to treat the inner planets as a group, the diversity of the planets and of their atmospheres is remarkable. The three processes of impacts, volcanism and tectonics are vital ingredients in the evolution of the planets. Our understanding also depends critically on the sources and dissipation of heat.

The origin of the atmospheres of the inner planets is an important topic in itself. It is now thought that the origin is due to outgassing of primitive material from which the planets were made. The subsequent evolution of the atmospheres depends on two major factors: the distance from the Sun (which controls radiation input) and the mass of the body (controlling first heat loss rate from the initial increase due to accretion, second heating rate due to radioactivity, and third atmospheric escape speed). The presence of life on Earth has also played an important role in determining atmospheric composition here.

**Figure 5.2.** Images from the three bodies (Mars, Venus, the Moon) visited by humankind or robot landers so far. Upper left is from Mars Pathfinder (courtesy of NASA JPL), lower left is from Venera 13 at Venus (courtesy of Russian Academy of Sciences), right shows Apollo 17 astronaut Harrison Schmitt (courtesy of NASA).

Mercury is a hot, heavily cratered planet which is difficult to observe because of its proximity to the Sun. Only one spacecraft, Mariner 10, has performed three fast flybys of the planet. Mercury is remarkable because of its high density, second only to the Earth. Another unexpected discovery from Mariner 10 was the presence of a strong magnetic field and a magnetosphere. It is likely that the planet has a larger iron-rich core in relation to its radius than the others. Mercury has an 'exosphere' rather than an atmosphere, since the pressure is so low that escape is as likely as a gas collision. An atmosphere was not retained because of the low mass (size) and high temperature (proximity to the Sun). The planet has an eccentric, inclined orbit. The rotation period is in a 2:3 ratio with its period of revolution around the Sun. This indicates that the slightly non-spherical shape of the planet was important during its formation. The surface of Mercury contains a well-preserved, little-disturbed cratering history. During the early bombardment in the first 0.8 billion years since formation, ending with a large impact which produced the 1300 km Caloris Basin feature, there was important tectonic activity. There is also evidence of significant shrinkage of the planet due to cooling, causing 'lobate scarp' structures; this shrinkage also occurred early in the planet's history.

The Venus–Earth–Mars trio are particularly important for us to understand because we know that life evolved at least on the Earth. In the case of Venus, the planet, although similar in size to the Earth, was closer to the Sun; water evaporated and caused a greenhouse effect, causing further evaporation of water and eventually a runaway greenhouse effect. Lack of surface water meant that fixing of evolved carbon dioxide in the rocks via absorption in the oceans as on Earth was impossible. Carbon dioxide evolved into the atmosphere over the four billion years since formation continues the greenhouse effect after the hydrogen (from water dissociation) has escaped to space and the oxygen has oxidised rocks. The surface temperature is now 750 K, the atmosphere is thick and supports sulphuric acid clouds, and our sister planet is completely inhospitable to life. The clouds are observed to rotate much faster (c. 4 days) in the equatorial regions than the planetary rotation rate (once per year); this 'super-rotation' is one of the aspects of the Venusian atmosphere that is not yet well understood.

In terms of the space programme, there have been missions to study the atmosphere (Mariner, Pioneer, Venera), map below the clouds using radar (Magellan), and landers (Venera) have transmitted pictures from the surface.

Mars, on the other hand, is smaller than the Earth and further from the Sun. Isotopic ratios between radiogenic $^{40}Ar$ and primordial $^{36}Ar$ indicate that only about 20 per cent of the gas in the rocks has been evolved into the atmosphere because of reduced tectonic activity (small size). Also due to the size, much of the early atmosphere was lost and the present atmospheric pressure is less than 1 per cent that of Earth. Substantial carbon dioxide ice deposits are present at the poles but there is not enough in the atmosphere to cause a greenhouse effect. Oxygen isotopic ratios show that there must be a source of oxygen, perhaps frozen sub-surface water, which increases the proportion of the lighter isotope, which would otherwise preferentially be lost to space, to the observed level. On the other hand, images from Viking and Mars Global Surveyor contain evidence that liquid water flowed on the Martian surface about 3.8 billion years ago. The present Mars is much colder, dryer and much less hospitable to life than it once was.

Some serious scientists propose that greenhouse gases could be introduced into the Martian atmosphere to warm the planet and release some of the trapped water and carbon dioxide, ultimately giving a hospitable environment for humans. This is an interesting idea in theory and a good target for computer simulations. In the opinion of this author, terraforming would be the ultimate in cosmic vandalism if implemented.

The Earth was at the right place in the Solar System and was the right size for life to evolve. The presence of liquid water on the surface meant that dissolved carbon dioxide could be fixed in rocks as carbonates, some of this is recycled due to volcanism. As life developed, photosynthesis became important, leading to the production of oxygen and the fixing of some of the carbon in the biomass. Enough oxygen in the atmosphere led to production of stratospheric ozone, which allowed the protection of land-based life forms from harmful EUV radiation.

The Earth is also the planet we know the most about. Looking at the Earth from space gave us a new perspective: an enhanced feeling that the Earth is special and indeed fragile. The average temperature of the Earth's surface is close to the triple point of water where solid, liquid and vapour may all exist. That is part of how we came to be here.

Our Moon is the first planetary satellite in terms of proximity to the Sun. Its density is much lower than the Earth's and there is effectively no atmosphere. The cratering record is therefore well preserved but the maria show that volcanic activity was important after the early bombardment and up to about 3.2 billion years ago. Despite intensive study by spacecraft (Luna, Ranger, Surveyor, Apollo, Clementine, Lunar Prospector) the origin of the Moon has not yet been determined from the competing theories (simultaneous formation, catastrophic impact on early Earth, capture).

The satellites of Mars, namely Phobos and Deimos, may be captured asteroids based on their physical characteristics. However, the understanding of the dynamics of their capture is by no means solved.

### Our planetary neighbours

Despite the proximity of our planetary neighbours and the many space missions which have explored them, many important questions remain. Why is Mercury's core so large? Might a catastrophic collision early in its life explain this and its orbital eccentricity and inclination? How is seismic activity affected? Is there ice at Mercury's poles? Why is there super-rotation in the Venusian atmosphere? What is the surface composition of Venus and Mercury? What is the geological history? How oxidised is the Venusian surface and what is the history of water in the Venusian atmosphere? What changes is humankind making to the Earth's climate and do these need to be ameliorated? What is the origin of the Moon? Do the Martian atmospheric loss rates to space support the models? Where is the water on Mars now? What is the history of other volatiles? Was there life on Mars? Could and should we terraform Mars?

## 5.6 The asteroids

In some sense the asteroids belong with the inner planets. Many of the asteroids occur in the main belt in between Mars and Jupiter. Some are in other orbits, including orbits which cross the Earth's path. A wide variation of eccentricities and inclinations of the orbits are also present. Spectral studies allow the classification of asteroids into several types: C-type, dark, rich in silicates and carbon, mainly outer main belt; S-type, rocky bodies, mainly inner main belt and Earth crossing; M-type, iron and nickel. A few other asteroids do not fit this scheme. It seems likely that asteroids are the remains of inner Solar System planetesimals rather than due to the destruction of a larger body. However, there have been collisions between some bodies since the early bombardment leading to fragmentation and other processing. Collisions with the Earth may have been important. In future, commercial mining of asteroids for minerals may become economically feasible.

### Asteroid missions

Why are asteroid types diverse? What is the composition? Which, if any, are planetesimals? How pristine? Do they contain interstellar grains from before the Solar System? Is there any water? What is their origin? Which asteroids do meteorites come from? Might they be a future source of raw materials?

## 5.7 The outer planets

The outer planets group contains the gas giants (Jupiter, Saturn, Uranus and Neptune) and the icy object Pluto. The gas giants are heavy enough and were cold enough when they were formed to retain the light gases hydrogen and helium from the solar nebula, and these constituents form most of the mass of the planets, reflecting the early composition. The visible disc for telescopes and space probes is ammonia- and water-based clouds in the atmosphere. At Jupiter, the largest planet and closest gas giant to the Sun, the cloud structure shows a banded and colourful structure caused by atmospheric circulation. The detailed cloud colours are not fully understood. There is no solid surface as such, but models of the internal structure of the gas giants show increasing pressure below the cloud tops; ultimately the pressure becomes so high that a metallic hydrogen layer forms at about 80 per cent and 50 per cent of the radius, respectively. Dynamo motions in this layer, assuming it must be liquid, drive the powerful planetary magnetic fields. A rocky/icy core is thought to be present at about 25 per cent of the planetary radius.

Jupiter rotates rapidly, providing some energy via the Coriolis force for atmospheric circulation. However, both Jupiter and Saturn have internal heat sources which mean that they emit 67 per cent and 78 per cent more energy than they receive from the Sun, respectively. This gives most of the energy for the atmosphere, but the origin of the internal heat source is not fully understood. Models indicate that helium precipitation within the metallic hydrogen core, in which helium is insoluble, may be responsible. There are also strong zonal (east–west) winds near the equator on Jupiter and Saturn, stronger on Saturn where they reach two-thirds of the speed of sound. Their origin is not fully understood. The planets also have important long-and short-lived atmospheric features, of which the most prominent is the Great Red Spot on Jupiter. This long-lived feature, seen for at least 300 years, is surprisingly stable and so far there is no adequate model to describe it. A similar spot feature appears on Neptune.

*In situ* results at Jupiter have recently been enhanced by data from the Galileo orbiter and probe. While the orbiter has discovered unexpected dipole magnetic fields in some of the Galilean satellites, the probe has sampled the atmospheric composition, winds, lightning and cloud structures at only one point, which turned out to be a non-typical location in the Jovian atmosphere. One of the discoveries was a lower-than-solar helium abundance which provides some support for the idea of helium precipitation in the metallic hydrogen layer; a similar conclusion was arrived at based on Voyager data at Saturn. Also there is less water than expected.

The gas giants each have important and fascinating moons. At Jupiter, Io has the only known active volcanoes other than Earth, providing sulphur-based gases for the Jovian magnetosphere; Europa may have a liquid-water ocean under its icy crust; Ganymede has its own 'magnetosphere within a magnetosphere'; and Callisto has a very old, cratered surface. Our knowledge of these has been revolutionised by *in situ* observation; before this only the albedos and orbital periods were known. At Saturn, Titan is a tantalising object, planet-like in structure and the only moon with a significant atmosphere – 1.5 times the Earth's pressure at the surface. However, its face was shrouded from Voyager's view by organic haze in its thick nitrogen–methane atmosphere. The atmosphere may hold clues about Earth's early atmosphere; there may be methane- or ethane-based precipitation systems; and the ionosphere forms a significant source for Saturn's magnetosphere. Cassini–Huygens will study Titan and some of Saturn's twenty or so icy satellites in detail starting in 2004. At Uranus, the moon Miranda graphically indicates the accretion theories as it appears to be made up of several different types of structure seen elsewhere. Also the moon system is out of the ecliptic because the spin axis of Uranus, at 98° inclination, is almost in the ecliptic itself. At Neptune, Triton is an icy satellite with a very thin atmosphere but it is in a retrograde orbit and is spiralling closer to Neptune; in tens of millions of years it may break up to produce spectacular rings. It may be similar in characteristics to Pluto and Charon.

Ring systems are present at all the gas giants but spectacularly so at

Saturn. Saturn's rings were discovered by Galileo, found to be separate from the planet by Huygens and found to have gaps by Cassini, who also suggested that they were composed of separate particles; this idea was mathematically proved two centuries later by James Clark Maxwell. Detailed exploration was begun by the flyby missions Pioneer and Voyager, which found remarkable structures including warps, grooves, braids, clumps, spokes, kinks, splits, resonances, and gaps. Whole new areas of Solar System dynamics were opened up, including the study of electromagnetic forces which may be important in spoke formation. The rings are less than a kilometre thick, as low as tens of metres in places and composed of billions of chunks of ice and dust ranging from micrometres to tens of metres in size. But the main question has not yet been satisfactorily answered: how did the rings form? Was it break-up of a smaller satellite or cometary capture?

And then there is Pluto, with its moon Charon. Following an elliptical, inclined orbit and currently the furthest planet, Pluto is an icy body rather than a gas giant. It may be closely related to, but larger than, the icy Kuiper belt objects, the outer Solar System planetesimals, and it may also be related to Triton. Much will be learned by the first spacecraft reconnaissance of the Pluto–Charon system. However, as Pluto goes towards aphelion its tenuous and interesting methane-based atmosphere will condense and become much less dense. In 2010–2020 it is expected that a rapid atmospheric collapse will occur. There is a good case for getting to Pluto as soon as possible and another excellent case for a visit near the next perihelion in 2237.

### Questions for the new millennium on outer planets and their satellites

What causes the cloud colours in the gas giants? Are the internal structure models correct? What causes the internal heat source? Why are the zonal winds so high on Jupiter and, particularly, Saturn? Why are atmospheric features, such as the Jovian Great Red Spot, so stable? Does Europa have water oceans and is life a possibility there? Will the Titan atmosphere evolve futher when the Sun becomes a red giant? Why is the Uranian spin axis so tilted and what are the effects? Can the study of Saturn's rings give us more information about the radiation–plasma–dust mixture in the early Solar System? What are the basic characteristics of Pluto–Charon?

## 5.8 The comets

Comets are the building blocks of the outer Solar System. Their formation was at the low temperatures prevalent in the primordial nebula and they retain volatile material from the early Solar System. They are relatively pristine bodies, making their study important. From the orbits of comets we find that some were formed in the Uranus–Neptune region but were expelled to form the spherical (and theoretical) Oort cloud with radius approximately 50 000 AU. The orbits of these distant members of the Solar System may be disturbed by passing stars. They may then plunge into the inner Solar System where their orbits have random inclinations. Others form the Kuiper belt just beyond Pluto's orbit; their inclinations are close to the ecliptic plane. Whatever their origin, comet nuclei are dirty snowballs which, when they near the Sun, emit gas and dust which form the plasma and dust tails. Halley's comet, for example, loses about a metre of material from the surface of the nucleus per orbit and has orbited the Sun about 1000 times, but activity varies significantly from comet to comet.

The space missions in the mid-1980s confirmed that comets have a distinct nucleus, measured gas, plasma and dust composition and led to an understanding of tail formation. The main surprises were the darkness of the nucleus and the jet activity rather than uniform gas and dust emission.

Cometary collisions played a key role in the inner Solar System. Collisions can still occur now, as shown by the Shoemaker–Levy 9 impact with Jupiter in 1994.

### Cometary missions

Given their importance in the early Solar System, what is the detailed composition of several comets? Can we bring an icy sample back to Earth for analysis? Is there an Oort cloud? What is the relation to planets? Might comets have brought volatiles to the inner planets?

## 5.9 The interplanetary medium and solar wind interactions

The aurorae and cometary tails, the only visible clues to the solar wind, have been observed for centuries. Occasionally the aurora borealis is seen as far south as Britain. Near the dawn of the last millennium, in 979 AD, when Ethelred (the Unready) was pronounced King, a possible occurrence was recorded in the *Anglo-Saxon Chronicle*:

In this year Ethelred was hallowed king at Kingston (Surrey) on the Sunday,
fourteen nights after Easter . . . In the same year was seen often-times a bloody
cloud, in the likeness of fire; and that was most apparent at midnight; and was
coloured in various ways. Then when it was about to dawn, it glided away.

Although it sounds ominous he reigned until 1016. As with many other
scientific phenomena, the beginning of understanding of the aurora had to
wait until the present era. Gilbert's ideas (in 1600) of the Earth as a magnet,
Halley's (1698 and 1700) magnetic maps of the Earth from ships and his
idea of the aurora being associated with the magnetic field, and George
Graham's 1722 observation of the motion of compass needles were important early contributions. More recently, Birkeland fired electrons ('cathode
rays') at a magnetised sphere in a vacuum in his terrella experiments in the
early 1900s, Appleton and others studied the ionosphere in the 1920s, and
Chapman laid the foundations of modern solar–terrestrial theory starting
in 1930; these all contributed to our current understanding of the solar–
terrestrial relationship. At the dawn of the space age, the first and completely unexpected scientific discovery from the space programme was of
the radiation belts of energetic charged particles trapped in the Earth's magnetic field.

Observations of comets in the 1950s had led to the idea of a constantly
blowing but gusty solar wind. The solar wind was confirmed by early space
probes but it was not until the mid-1980s that the comet–solar wind interaction was understood and backed up by *in situ* data. Solar–terrestrial
research, and solar wind interaction with other bodies, remain active areas
of research into the new millennium. One of the drivers for understanding
is the effect that violent solar activity can have on the electromagnetic
environment of the Earth and on humankind's technological systems. I
have been privileged to be part of some of the exciting exploratory space
missions in these fields.

In its present state, the Sun emits about a million tonnes per second of
material in the form of plasma. At this rate it would take $10^{14}$ years to dis-
integrate; well before then, in about 5 billion years, the hydrogen fusion
fuel will be exhausted and the Sun will become a red giant. In the mean-
time, the Earth, other planets and comets are all bathed in the solar wind.

The solar wind is highly conducting such that, to a good approximation, the magnetic field is frozen into the flow. Plasma in the solar corona
is hot enough to escape the Sun's gravity along the magnetic field where

**Figure 5.3.** The solar source and terrestrial effect of the solar wind. Left, a SOHO image from 26 February 1998, in the light of eight and nine times ionised iron at 17.1 nm, reveals the magnetic structure in the plasma near to the Sun at a million degrees (courtesy of EIT team/ESA/NASA). Centre, an eclipse image from the same date showing the corona (copyright Fred Espenak). The eclipse occurred soon after sunspot minimum; coronal holes, characterised by open magnetic field lines, are visible. Right, an ultraviolet image of the Earth from the Dynamics Explorer satellite shows the aurora from space (courtesy of University of Iowa).

motions are unrestrained, and it can be shown that the solar wind becomes supersonic within a few solar radii. Beyond this, the speed is almost constant for a particular element of solar wind but the value can vary between about 300 and 800 km s$^{-1}$. As plasma expands radially through the Solar System it drags the solar magnetic field along, but this forms a spiral pattern in space due to the solar rotation. By Earth's orbit the density is about 5 cm$^{-3}$ but variable and disturbed by coronal mass ejections, shocks and discontinuities. As the solar cycle waxes and wanes the source region of the solar wind changes, in particular the coronal holes get smaller (increased magnetic complexity) and larger (decreased complexity) respectively. The electrodynamic environment of Solar System bodies is extremely variable on timescales of seconds (ion rotation around the magnetic field) to 11–22 years (solar cycle).

The extent of the solar wind reaches well beyond the planets. Ultimately a heliopause is required where the solar wind pressure balances that of the local interstellar medium (LISM), at about 150 AU on the upstream side. Before that, at about 100 AU, a terminal shock slows the solar wind from supersonic inside to subsonic outside, and upstream of the heliopause a bow shock may form if the LISM motion is supersonic. The terminal shock, heliopause, and bow shock are all hypothetical, as the Voyager spacecraft, the most distant artificial object, has not yet crossed these boundaries. However, the inner heliosphere is becoming better understood from several spacecraft, notably Ulysses which is measuring the structure out of the ecliptic plane in the Sun–Jupiter region for the first time.

The interaction of the solar wind with an obstacle depends critically on the obstacle itself: its state of magnetisation, its conductivity, and whether it has an atmosphere. We will consider two main types of object: a magnetised planet such as Earth and an unmagnetised object such as a comet.

Other objects include interesting features of both extremes. Mars and Venus are unmagnetised but have some cometary features. Io is conducting, is within the subsonic Jovian magnetosphere, produces a plasma torus, drives huge field-aligned currents to Jovian auroral regions causing light emission there, and supports Alfvèn wings. Titan has a dense ionosphere, it is usually in the subsonic magnetosphere of Saturn but is sometimes in the solar wind, field line draping seems to occur as at comets.

## 5.9.1 Magnetised planet interaction

The discussion concentrates on the Earth but is also relevant to Mercury, Jupiter, Saturn, Uranus and Neptune; differences are caused by different magnetic dipole strengths and orientations, spin rates, and particle sources such as moons and ionospheres, and sinks such as rings. In the case of Mercury, there is no ionosphere, so that at present we do not understand how the electrical currents close.

Magnetised plasmas do not mix. As the solar wind approaches the Earth, a current sheet, the magnetopause, is set up to separate the regions of solar and planetary magnetic field. The solar-wind particle pressure outside balances the magnetic pressure inside (these are the dominant pressure components), so that the magnetopause is compressed on the day side and extended like a comet tail on the night side. This simple model was formulated by Chapman and Ferraro early last century and gives a good prediction for the magnetopause location. Outside the magnetopause a bow shock stands in the supersonic solar wind flow. The nature of this collisionless shock changes with magnetic field orientation as, to some extent, the magnetic field plays the role that collisions play in a fluid shock.

However, the real situation is not this simple. Shocked solar wind particles can penetrate the magnetopause directly via the funnel-shaped cusp regions on the day side to cause day side aurora. Also, if the solar-wind

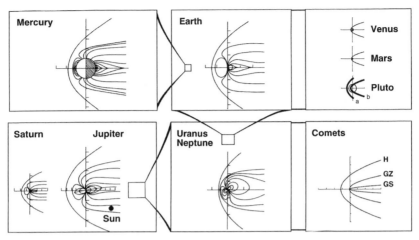

**Figure 5.4.** Comparison of the sizes of magnetospheres and non-magnetic interactions with the solar wind (adapted from Kivelson and Russell, 1995). The two panels on the right illustrate the bow shocks of non-magnetic objects. In the Pluto panel, Charon's orbit is shown as a circle, and the anticipated shock locations are shown when Pluto is (a) closest to the Sun and (b) furthest so that the atmosphere is more tenuous. Bow shock locations of the three comets visited so far by spacecraft are shown as H (Halley – Giotto), GZ (Giacobini – Zinner – ICE) and GS (Grigg – Skjellerup – Giotto).

magnetic field and the northward directed terrestrial magnetic field (the magnetic dipole of Earth is northward in the Southern Hemisphere at present) are oppositely directed, then it was realised by Dungey that our nice fluid model is not enough. The explosive process of magnetic reconnection takes place at just the distance scale where the fluid approximation breaks down. This causes solar wind field lines to be connected to terrestrial ones and these are dragged over the polar cap like peeling a banana. Ultimately this leads to a buildup of magnetic energy in the tail, further reconnection in the deep tail, and perhaps important trigger processes nearer to Earth. The effect is that some plasma is shot down the tail and some towards Earth, causing night side aurora. The reconnection process, and the electric field across the tail caused by the motion of the solar wind relative to the magnetised Earth, sets up a convection system.

Another convection system is caused by plasma corotating with the planet. The atmosphere and ionosphere corotate with the Earth, as does the inner part of the magnetosphere – the plasmasphere. The overall circulation in the Earth's magnetosphere is given by the sum of the corotation electric field and the convection electric field. As we can see, the convection part of this is extremely dynamic and is still the focus of intense research.

Measurements from ground and space are used together to diagnose the near-Earth environment. Despite the success of early satellites in mapping the various regions, one of the major difficulties of space measurements has been the use of single or dual satellites and the consequent space/time ambiguity. If one or two satellites see a signal due to a boundary, how can we know if the boundary has moved over the satellite or vice versa? The only way to resolve this is with more satellites. At present this is possible on the large scale with the ISTP (International Solar-Terrestrial Physics) fleet. Satellites upstream of the Earth monitor the Sun and solar wind (SOHO, ACE, Wind), while satellites in the magnetosphere (Polar, Interball, FAST and others) and in the tail (Geotail) monitor the overall effects.

This combination has proved extremely useful in following coronal mass ejections from the Sun, through the interplanetary medium and to the Earth. In one case the consequent increase of radiation belt particles (via an as-yet unknown process) may have caused a commercial telecommunications satellite to fail. This was the first time that 'space weather' had been measured while causing catastrophic effects. As electronics integration and our dependence on technology increases these effects will become more important still. Effects are felt on satellites and in electricity cables and oil pipelines where currents are induced.

On a small scale the Cluster mission will be the first to make progress here. This is a group of four spacecraft launched in 2000 following an abortive attempt on the first Ariane 5 in 1996. They are flying in tight formation, starting a few hundred kilometres apart, in a polar orbit through the magnetosphere. At one part of the orbit the spacecraft will be at the corners of a tetrahedron. With two spacecraft we can measure along a line in one dimension, with three we measure a two-dimensional plane, but with four spacecraft we can measure in three dimensions. For the first time the space-time ambiguity will be resolved and it will be possible to measure vector gradients. This gives the exciting prospect of being able to measure the parameters in Maxwell's equations of electromagnetism directly and to do some real plasma physics with the results. The reconnection process, cusp entry processes and tail dynamics will be directly measured.

At Saturn, the Cassini mission will orbit the planet for four years. This will give us more information about the similarities and differences to Earth's magnetosphere. Saturn's magnetosphere is controlled by the solar wind but corotation is more important, the magnetosphere is bigger so timescales are longer, and the rings and moons make important differences. This will be one mission to contribute to studies of comparative magnetospheres and will help to tell us more about our own.

For the future, even the large-scale ISTP constellation and the small-scale three-dimensional Cluster measurements still lead to a significant undersampling in near-Earth space. This is now starting to be addressed by two techniques: magnetospheric imaging and multi-spacecraft (30+) missions. These will be important techniques in the coming decades.

*Questions on interplanetary medium and magnetized planet interaction*
How do solar and interplanetary events relate? Will it be possible to forecast events at Earth, particularly the sign of the interplanetary magnetic field? Is there a terminal shock, heliopause and bow shock far outside the planets? What triggers reconnection? What causes energy release in the tail? What do different timescales and other differences at the planets teach us about Earth's magnetosphere? How are radiation belt particles accelerated? Can we develop better protection techniques for our satellites? How do field-aligned currents in Mercury's magnetosphere close? Is it important that the planet occupies much of the inner magnetosphere? How does this magnetosphere work? How do the ecliptic plane orientation of the Uranus geographic axis, the 60° tilt of the magnetic axis with respect to this, and the rapid rotation rate affect the Uranian magnetosphere? How is the planetary magnetic field produced at Mercury, Uranus and Neptune? How important to the survival of life on Earth was the protection from solar wind and cosmic radiation afforded by Earth's magnetic field?

## 5.9.2 Comet–solar wind interaction

A comet interacts with the solar wind in quite a different way from a magnetised planet. The cometary nucleus is not magnetised and is a 'dirty snowball'; in the case of comet Halley the nucleus is some 15 km by 8 km as determined by the Giotto spacecraft. When the nucleus warms near to the Sun, gas (mainly water vapour) sublimes away at about $1 \, km \, s^{-1}$, carrying dust with it. Comet Halley, for example, produces about $20 \, t \, s^{-1}$ while Grigg–Skjellerup, Giotto's weaker target, produced only $200 \, kg \, s^{-1}$ at the encounter time. The gas may ionise in sunlight or by charge exchange with the solar wind, on a timescale of a week or so. When ionised, the new-born ion interacts with the electric and magnetic field in the solar wind.

The new ion is first accelerated along the electric field and then gyrates around the magnetic field. The motion of the particle is a cycloid, the same path taken by the valve on a bicycle wheel as it moves. The motion is well known to plasma physicists as 'E cross B drift'; the speed of the centre of the motion is given by the ratio of the electric to magnetic fields. The motion in real space is equivalent to a ring in velocity space, centred on the drift speed in a direction along the magnetic field.

The ring causes plasma instabilities which excite Alfvèn waves moving predominantly upstream along the magnetic field. We have been able to show both experimentally and theoretically that, due to energy conservation, energy from the particles is given to the waves, causing the particle distributions to follow bispherical shells in velocity space, centred on the upstream and downstream waves. This can cause particle acceleration as well as deceleration and, contrary to expectation, comets are good particle accelerators.

The solar wind is slowed due to the added mass (it is 'mass-loaded'). This leads to draping of the magnetic field around the comet as predicted by Alfvèn as the plasma is frozen to the flow. If the slowing proceeds rapidly enough a bow shock is formed in the flow. This was observed at the three comets visited by spacecraft so far, although in some cases the boundary was a 'wave' rather than a shock. Due to the cometary ions, whose gyration radius is much larger than that of the solar wind particles, cometary bow shocks are the most complex in the Solar System.

The regions inside the bow shock gave several surprises and boundaries not predicted by models. Nearest to the nucleus, however, a predicted boundary appeared where the magnetic field plummeted to zero: the cavity boundary. The number of cometary ions was so high here that the magnetic field was excluded completely.

Following the spacecraft encounters we have a detailed understanding of some of the physics at work in this interaction. The obstacle to the solar wind is very diffuse and dependent on the outgassing rate of the nucleus and its position in the Solar System. At small comets we even discovered non-gyrotropic ring distributions.

Since the encounters, an exciting discovery has been made of X-rays produced by comets, and it appears that the best explanation is due to the decay of excited states formed by charge exchange between heavy solar wind ions and the cometary ions.

### Questions on solar wind–comet interaction

How permanent are the various plasma boundaries? Exactly how is the solar wind slowed by the mass loading? Is particle acceleration from the diffusing cometary ions enough or are other mechanisms needed? How does the cometary tail form, how does the magnetic cavity connect to it and what is the importance of tail rays seen in remote observations? Could we fly along a comet's tail to better understand it? What can comets tell us about instabilities in Earth-bound fusion machines or about astrophysical phenomena such as supernova explosions?

## 5.10  Future exploration

During our review we identified some important questions to be answered for each class of body or region in the Solar System. In order to answer these questions, and others which will arise as some are answered, it will be necessary to continue space exploration far into the new millennium. Remote sensing techniques from the ground or from Earth orbit are unlikely to have sufficient resolution, the ability to penetrate clouds at the target, or the ability to see the far side of objects. In addition, the *in situ* measurements of plasma, dust, composition, and direct sampling cannot be done remotely at all. Most of the questions and studies highlighted play an important part in answering why humankind has evolved here. Some are directly related to the possible existence of life elsewhere. Answering the questions is thus of important cultural value as well as purely scientific curiosity.

Table 5.1 shows a list of the past missions and future missions approved over the next decade or so. The natural sequence of Solar System missions involves four stages: (1) initial reconnaisance by flyby; (2) detailed study by orbiter; (3) direct measurement of atmosphere or surface via entry probe; and (4) sample return. The stage we have reached for each body is also shown in Table 5.1.

The approved programme includes a mission to pursue the exploration of Mercury; an important series of missions to Mars culminating in *in situ* searches for life and sample returns to Earth; the exploration of two possible future sites for natural life, namely Europa and Titan; an in-depth

Table 5.1. In situ *Solar System exploration*

| Object | Past missions | Stage | Future missions (approved) |
|---|---|---|---|
| Mercury | Mariner 10 | 1 | ESA cornerstone Messenger |
| Venus | Mariner, Pioneer Venus, Venera, Vega, Magellan | 3 | — |
| Earth | Many | n/a | Many |
| Moon | Luna, Ranger, Surveyor, Zond, Apollo, Clementine, Lunar Prospector | 4 | Lunar-A, Selene, SMART-1 |
| Mars | Mars, Mariner, Viking, Phobos, Pathfinder, Global Surveyor | 3 | Mars Odyssey 01, Mars Exploration Rovers 03, Mars Reconnaisance Orbiter 05, Nozomi, Mars Express |
| Jupiter | Pioneer, Voyager, Galileo, Ulysses | 3 | Europa Orbiter |
| Saturn | Pioneer, Voyager | 1 | Cassini–Huygens |
| Uranus | Voyager | 1 | — |
| Neptune | Voyager | 1 | — |
| Pluto | — | 0 | Pluto–Kuiper Belt Mission |
| Asteroids | Galileo, NEAR | 1 | DS 1, Muses-C |
| Comets | ICE, Sakigake, Suisei, VEGA, Giotto | 1 | Stardust, Rosetta, CONTOUR, Deep Impact |
| Sun + i/p medium | WIND, ACE, ISEE, AMPTE SMM, Yohkoh, SOHO, TRACE, IMAGE | n/a | Cluster, Genesis, Solar-B, Solar Probe, STEREO |

exploration of Saturn's system; the first reconnaisance of the Pluto system; asteroid and comet landers and sample return missions; solar wind sample return; and the first near-Sun (four solar radii) flyby. This is an exciting and vibrant programme. But is the current drive, forced by budgetary necessity, for 'small, fast, cheap' robotic missions, working? One is tempted to answer yes, as every mission in the table is answering important questions and new missions can be proposed rapidly in response to new discoveries. But the payload carried by each of the missions is limited and only a few questions can be studied by each mission. Also the hiatus in NASA's Mars exploration programme caused by the failure of both Mars 98 (Climate Orbiter and Polar Lander) missions raises important questions on how fast and cheap missions can be and still achieve success.

However, some of the missions (Cassini in particular) are not of the smaller, faster, cheaper type. The strength of the Cassini mission is in its multidisciplinary approach: we will only understand the complexities of Titan's atmosphere by using several techniques, the *in situ* Huygens probe and several measurements from the orbiter. The probe will only measure at one place and, as seen at Jupiter, it may not be representative. We must be careful, therefore, to include some larger missions with a balanced payload in the future programme. For possible missions to the outer planets in particular it would be better to send larger, multidisciplinary missions. Another approach could be multispacecraft, multipurpose missions. The smaller, faster, cheaper philosophy only really seems relevant in the inner Solar System.

One problem in Solar System exploration is the time taken to get to the target. Cassini took eight years to build and the flight time to Saturn is another seven. While some opportunistic science is possible, in this case at Venus, Earth, Jupiter and in the distant solar wind, it will take a significant proportion of the careers of the scientists involved for Cassini to arrive – those that are not retired first. To get a reasonable payload to Pluto would take much longer than the seven years foreseen for the small mission with severely limited payload that is the Pluto–Kuiper Express.

If missions within the Solar System take a long time to reach the destination, is it realistic to consider missions beyond this? Using current spacecraft technology, assuming the same speed as Voyager (3.3 AU per year), missions to the heliopause at 150 AU would take over 50 years, the Oort cloud at 50000 AU would take at least 15000 years, and to Proxima Centauri, our nearest star at 4.2 light years away, would take about 80000 years. It is clear that better trajectories and advanced propulsion systems would be needed for the more accessible missions, and remote sensing is a better technique for the stars. Within and closer to the Solar System, ion propulsion or inner Solar System light-driven designs may be appropriate. Another problem for these remote missions will be the provision of electrical power so far away from the Sun. The only solution appears to be nuclear power.

As we find out more about each object or region, further questions will be raised. It will also be necessary to explore further afield. Even if firm evidence is found for past life on Mars we will need to understand how common the occurrence of life is in the universe, hence the search for extra-solar planets. We need to understand our own Solar System properly before extrapolation is possible.

We are making good use of space for peaceful meteorological, communications and positioning reasons on Earth. Uses of space may also reach further into the Solar System in the new millennium: for example, there may be economic sense in asteroid mining. Another example is mining the Moon for helium. Because the Moon is unmagnetised, the solar wind can impact it directly. The magnetic field diffuses relatively rapidly but the particles are buried in the regolith. Over billions of years, enough $^3$He may have been buried to make mining of this isotope worthwhile for use in future fusion reactors on Earth. Another possible, but in my view unpalatable, application is planetary engineering or terraforming.

One controversial issue is the direct involvement of humankind in *in situ* exploration. So far only a handful of people have travelled beyond 400 km and only a few hundred beyond 100 km from the Earth's surface. However impressive the missions and brave the people involved, the reasons for sending them are to do with politics and public relations. These were associated with national pride during the cold war, and science was an afterthought. Lunar samples were returned by Russian robots as well as by the Apollo programme, although admittedly much less mass. Humankind as a whole can be proud of the achievements but it seems unlikely that the same political will or economic priority will exist to propel people far beyond low Earth orbit.

On the other hand, robots have successfully explored all the planets except Pluto, and the furthest man-made object, Voyager 1, is over 70 AU away and counting. Do we need the encumbrance of life-support systems? Can we accept human error in space? Would space travellers be adequately protected from radiation away from the protection of our magnetic field except in an extremely thick-walled spacecraft? Is there the economic or political will or necessity to support expensive manned exploration? If computer and virtual reality techniques continue to accelerate in development do we need to send people? From a purely scientific perspective the answer is no.

Another obstacle for any kind of mission is launch cost. This is the prime reason that manned exploration is likely to stay in Earth's vicinity, and is also a severe limitation on robotic exploration. Launches more frequent than the present rate will have to await technological developments on re-usable rockets, aerospike engines and perhaps air-breathing engines. Studies are currently underway in the United States with the X-33 demonstrator being built, and it seems likely that this technology will become usable in the next decade or so.

Robotic space exploration should be continued as rapidly as possible. This tool, which has become available within the last half century, has already proved its worth and should be used for important scientific purposes. Robotic exploration is much better value for money.

We may speculate on possible targets for robotic missions for the new millennium as follows: constant monitoring of the Sun and solar wind as part of an integrated space weather forecasting system; constant monitoring of weather on Mars, Venus and Jupiter to improve models; detailed explorations of Mercury and Pluto; return to sites of earlier exploration with better instruments, new ideas and atmospheric probes in several places; sample return from nearby Solar System bodies following detailed mapping and *in situ* composition measurements; exploration of Uranus and its extraordinary magnetosphere; explorations of outer planetary moons; investigation of feasibility of asteroid mining; terminal shock, heliopause, heliospheric bow shock exploration; Oort cloud exploration; investigation of feasibility of sending spacecraft to nearby stars and planetary systems.

## 5.11 Other solar systems and life elsewhere

One of the most fundamental questions facing humankind in the new millennium is: has there been or could there be life elsewhere? The question has been haunting us for at least two millennia, but the answer is now closer given recent developments in technology which are overcoming the vast difficulties for our observations. The answer will have profound and exciting implications for scientific, cultural, philosophical and religious thinking. Most scientists in the field now believe there was insufficient evidence for NASA's 1996 announcement of early life on Mars. Nevertheless, the announcement of the result was introduced by the President of the United States, generated huge interest and drew comment from leaders in many fields in addition to science. This illustrates the importance of the answer to humankind. Without life elsewhere, we feel alone and isolated in the universe. Whether life has or has not evolved elsewhere yet, we would like to know why.

As we look into the night sky, at the 3000 or so visible stars, it is natural for us to speculate on life elsewhere. The knowledge that our galaxy alone is some 100000 light years across, contains some 200 billion stars, and that the universe contains some ten times as many galaxies as the number of stars in our Milky Way, makes the presence of life elsewhere seem possible and indeed likely. Attempts have been made to quantify this, and the Drake equation is still the best way of writing down the number of contemporaneous civilisations in our galaxy as the product of seven terms: the formation rate of stars, the fraction of those with planets, the number per Solar System that are habitable, the fraction where life has started, the fraction with intelligence, the fraction with the technology and will to communicate, and the lifetime of that civilisation. Reasonable estimates for all of these parameters from astronomy, biology and sociology lead to values between zero and millions of civilisations within our own galaxy. Although we only know the first term with any accuracy as yet, my guess would be towards the higher end. Without firm evidence it is impossible that we will ever be able to determine most of these terms accurately enough, and the final answer will remain indeterminate by this approach. Unfortunately, it seems that we are in the realm of speculation rather than science, creating a ripe area for science fiction.

Attempts to detect life elsewhere have included passive searches for organised signals and active approaches including the transmission of powerful microwave bursts and attaching plaques to the Pioneer and Voyager spacecraft leaving the Solar System. Despite recently enlisting the assistance of much computing power around the world, the passive search approach feels forlorn, but is still worth a try. As for our own electromagnetic transmissions, they have reached a very small part of our galaxy in the 100 years since radio was invented and in the 26 years since the Arecibo telescope was used to shout to the cosmos 'we are here'. We cannot get around the inverse square law for intensity of electromagnetic waves, making detection of our signals highly challenging – and in any case it would take a significant time to get an answer. Discounting an enormous stroke of luck, a more scientific and systematic approach is likely to be via the detection and analysis of extra-solar planets by remote sensing. The technology for this is just becoming available.

In the last few years we have seen the first tantalising evidence for other solar systems than our own, in different stages of formation. In 1984, the star Beta Pictoris was seen to be surrounded by a cloud of gas and dust reminiscent of early Solar System models. Other examples of dust–gas discs around stars have been detected recently using the Hubble Space Telescope. Have planetesimals, or comets, formed near these stars? Are observed structures in the cloud evidence for forming planets? We should begin to be able to answer these questions soon, and we should find many other examples.

So far, planet-hunters including Geoffrey Marcy have found firm evidence for over 30 planets around Sun-like stars, and several additional candidates. The main technique used for this is to detect the anomalous motion, or 'wobbling', of the companion star, using large ground-based telescopes, although in one reported case to date an occultation, or reduction of the companion star's light due to the object crossing its disc, was detected. The planets are all inferred to be massive gas giants, most are several times heavier than Jupiter and the smallest so far is the size of Saturn. All orbit closer to the star than the giant planets in our own Solar System, the main hypothesis for this being that they form further out and lose orbital energy via friction with the remaining gas and dust. Multiple large gas planets around another star have also been inferred from observations. One problem in the identification of any extra-solar planets is to distinguish them from 'brown dwarf' companions, failed stars which were too light for fusion to start. Telescope technology is advancing, and dedicated observational space missions using interferometry have been proposed and will be flown within the next decade. We must choose the most promising stars to observe, starting with stars like our own Sun. We can expect that the present catalogue of 30 will increase soon and significantly; the observable planetary size will decrease, becoming closer to the Earth's size;

accurate distinction techniques between planets and brown dwarves will be developed; and statistics will be built up on size distributions and orbits. We will ultimately know whether our own Solar System is unique.

Given the, in my opinion, likely existence of small rocky planets elsewhere, and the fact that we are most unlikely to have the technology to make *in situ* measurements, we will need to establish methods for detecting the presence of life on these objects using remote sensing. Using data from the Galileo spacecraft during its Earth swingbys, Carl Sagan and colleagues detected the presence of life on Earth using atmospheric analysis and radio signal detection. The challenge is enormously greater for planets of remote stars. However, we may speculate that spectroscopic observations of ozone, oxygen and methane may hold the key, at least to finding clues for life as we know it. But it is unlikely we will be able to prove conclusively that life exists there. There is an optimistic view, shared by many scientists including myself. Surely amongst the many planets elsewhere there must be at least some which have similar conditions to the Earth? The accepted critical planetary properties for life to emerge are size, composition, stellar heat input and age. To this list we should add the presence of a magnetic field to protect against stellar winds and radiation.

Within our own Solar System, where *in situ* exploration is possible, it is certainly worth searching for evidence of life on Mars during its early, warmer, wetter history. Several space missions are planned for this. In addition, two other possible sites for life within our own Solar System should be reiterated here and explored. First, Europa may be a possible present site for life, in its liquid water ocean underneath an icy crust, deep enough to be shielded from Jupiter's powerful radiation belts. Second, Titan is a potential future site for life when the Sun exhausts its hydrogen fuel, becomes a red giant and warms the outer Solar System. As well as looking for tangible clues within our own Solar System it is clear that we must compare with other solar systems and broaden the search.

## 5.12 Conclusions

In summary, there are many exciting and challenging ways we can explore Solar Systems in the new millennium: our own with *in situ* studies and beyond with remote sensing. The answers to be gained are fundamental to a better understanding of our place in the universe.

## 5.13  Further reading

Lewis, J. S. 1997 *Physics and chemistry of the Solar System*, London: Academic press.

Beatty, J. K., Petersen, C. C. & Chaikin, A. 1999 *The new Solar System*, Cambridge: Cambridge University Press.

Stern, S. A. 1992 The Pluto–Charon system, *Ann. Rev. Astron. Astrophys.* **30**, 185–233.

Kivelson, M. G. & Russell, C. T. (eds.) 1995 *Introduction to space physics*, Cambridge: Cambridge University Press.

# 6
# Unveiling the face of the Moon

## Sarah K. Dunkin[1,2] and David J. Heather[3]

1 Space Science and Technology Dept., Rutherford Appleton Laboratory, Chilton, Didcot, OX11 0QX, UK
2 Dept. of Geological Sciences, University College London, Gower St., London WC1E 6BT, UK
3 ESTEC, SCI-SO, Keperlaan 1, Postbus 299, 2200 AG Noordwijk, The Netherlands

The Moon is considered by many planetary scientists to be the most important extra-terrestrial body in the Solar System. Preserved within its outer layers is an accessible record of the changing conditions in the inner Solar System over billions of years, a feature that is unique to the Moon as similar records have long since been destroyed or obscured on the other, more active inner planets. Because of this, the Moon is the key to unlocking the secrets of the origin and evolution of the Earth and inner Solar System. The long term goal of lunar science is to use that key, describing the evolution of the Moon from its creation to the present day and relating this to other planetary bodies where possible. Despite its proximity to Earth, the Moon has proved to be a difficult object to analyse. For one, it reveals only one face (the nearside) to us, keeping the other permanently turned away, and it was not until the advent of the space age in the late 1950s and early 1960s that we got our first views of the lunar farside. The two faces are surprisingly different (Figure 6.1), the nearside showing both maria (dark, smooth lava plains) and the more ancient highlands or 'terra' (bright mountainous areas), while the farside is almost exclusively highland in nature. The reason for this difference is still debated by lunar scientists. Even though the Moon is considered by planetary scientists to be a relatively 'simple' body, after centuries of telescopic observation and decades of spacecraft exploration we are still unable to describe some of the most fundamental aspects of its history. The close of the Apollo era left us tantalisingly close to revealing some of the greatest mysteries of the early

**Figure 6.1.** The two faces of the Moon. The nearside, to the left of the image, is distinctive by its dark, smooth areas known as maria. These are ancient lava flows that once flooded large depressions hundreds of kilometres in diameter around 3–4 billion years ago. A stark contrast to these dark areas is the bright highlands which dominate the farside of the Moon on the right. The mountainous highlands are all that is left of the original crust formed on the Moon over four billion years ago, and show the scars of bombardment by meteorites over the aeons. Why the farside is so different from the nearside remains to be discovered, and is a much debated point in lunar science. (Images courtesy of NASA.)

Solar System, but still our satellite refused to surrender its most precious secrets. However, recent lunar missions have sparked a period of renewed interest in the Moon, especially with the prospect of water ice being found at the poles. With the exciting results returned from these missions, a strong case for future exploration can be built, and we may at last be able to address key points which will lead us one step closer to understanding our enigmatic satellite.

## 6.1 The Apollo years

At the peak of lunar exploration, from 1959 to 1976, a flotilla of manned and unmanned spacecraft was sent to the Moon. Although the original motives for these missions were political (with the 'space race' between America and the USSR in full flow), a vast amount of science was also carried out as a necessity, and photographs, remote sensing data, even samples of lunar rocks and soil arrived on Earth for analysis. Humankind took its first steps on another world during this period, a historic event in its own right, but from a scientific perspective, the significance of putting a human being on the Moon cannot be understated. Not only did the six manned Apollo landings return to Earth with over 380 kg of rock and soil samples, but without the drive to achieve those landings, the precursor unmanned reconnaissance missions would never have flown, depriving us of an invaluable photographic dataset, still in use today.

Prior to the return of these data, there were three main theories relating to the origin of the Moon. First, the fission hypothesis, in which a fast spinning Earth threw off material to form the Moon; second, the co-accretion hypothesis, with the Earth and Moon forming together from the solar nebula (the cloud of gas and dust from which Sun and all of the major planets originated); and third, the capture hypothesis, where the Moon formed in a separate part of the Solar System and strayed too close to the Earth, being captured and held in orbit by the Earth's gravity. Each of these scenarios had positive and negative points and it was hoped that the returned lunar samples and other data would help to choose between them. However, using both the sample data and computational calculations, it soon became clear that all three theories had their problems. The capture hypothesis is physically unlikely, and the Apollo samples showed the oxygen chemistry on the Moon to exhibit the same trends as on Earth, a strong argument for the two bodies having formed in the same part of the

Solar System. This would support the fission and co-accretion models. However, the Earth–Moon system does not appear to have enough rotational energy for the Earth to have flung material off in the manner required for the fission model, and the co-accretion theory is unlikely due to chemical differences between the two bodies, particularly differences in iron and volatile content. It was not until the first tentative estimates of the bulk composition of the Moon were presented that a possible solution to the problem was suggested. Initial measurements showed the Moon to have a similar composition to the Earth's upper mantle, the molten layer beneath the solid crust of the Earth. This led to a fourth theory being put forward: the 'giant impact hypothesis', which suggests that that the Earth was struck by a body the size of Mars very early in the history of the Solar System, throwing material off that later collected together to form the Moon. Although this theory is now generally accepted, it too has its problems, and it cannot be claimed that we understand this part of lunar history well at all.

The data from the early years of lunar exploration also led to a drastic re-appraisal of theories relating to the evolution of the Moon. The samples showed a stark contrast in composition between the rocks from highland regions and those from the maria. Mare samples were found to be denser than the highlands and they showed an important difference in trace element composition. It was primarily these facts that provided the foundations upon which the currently favoured theory of lunar evolution is built. This theory, known as the 'magma ocean hypothesis' suggests that early in the Moon's history its upper layers were molten to a depth of several hundred kilometres. During this period, the lighter material floated to the surface, while the denser minerals sank (or rather, did not float). The lighter material then solidified to form the bright highland crust, and the denser material, which would have remained molten for some time, later erupted onto the surface as lava through cracks in the crust to form the maria. This theory has survived since the 1970s, but the details remain unclear, and there is still some way to go before unified theories for both the origin and evolution of the Moon are developed. Both the 'giant impact' and 'magma ocean' hypotheses are based on an estimate of the average composition of the Moon which itself may be flawed. The lunar samples and remote sensing data from the Apollo era, from which the average lunar composition was estimated, were restricted primarily to the equatorial regions of the nearside, and no compositional information was available at all for the lunar farside. Even 20 years after the last Apollo mission returned to the Earth, we lacked the one thing which is crucial to the understanding of the Moon: a global compositional dataset.

## 6.2  Our return to the Moon

A global view of the Moon was a long time coming. There now exists an entire generation who saw no major lunar mission until the recent unmanned Clementine and Lunar Prospector spacecraft were successfully launched in the 1990s. In the years after the last lunar missions of the 1970s, much work was done using the lunar samples to calibrate techniques to study the Moon remotely, particularly in the field of reflectance spectroscopy, involving the study of light reflected from a surface. All minerals absorb light at different wavelengths and reflect a unique 'spectral signature'. Therefore the composition of a soil or rock can be inferred by observing the fingerprints of its constituent minerals in its reflected spectrum. Multispectral techniques can make observations across several areas in a spectrum, thereby increasing the number of minerals that can be identified. Each zone observed within a given spectrum is known as a waveband, and the more wavebands that are covered, the higher the spectral resolution of the measurement and the more minerals can be recognised. The use of reflectance spectroscopy from Earth-based telescopes clearly demonstrated its potential as a powerful analytical tool, but was restricted by its poor spatial resolution and in its coverage (since the farside cannot be seen from Earth). To get the global compositional information needed to test the hypothesis of evolution and origin, a multispectral device would have to be flown on an orbiting spacecraft.

The Jupiter-bound Galileo spacecraft provided an idea of the science to come when, in 1990 and 1992, it completed two lunar flybys and tested its multispectral cameras by looking at the Moon. The resulting data covered just seven wavebands, was of a poor spatial resolution (several km/pixel) and surveyed only a portion of the lunar surface. In spite of these limitations, it was the first multispectral reflectance data of the Moon returned by a spacecraft and provided an abundance of useful information for lunar scientists.

Two years later, in January 1994, the US Department of Defense launched the Clementine spacecraft, carrying a scientific payload provided by NASA that included a multispectral camera. Its primary mission was to complete a flyby of the near-Earth asteroid Geographos, but it would visit the Moon en route. Two months in lunar orbit saw the return of over one million digital images at resolutions between 80 and 330 m/pixel, and, although the spectral resolution was poor in comparison to Earth-based observations, the dataset was global. For the first time, we had a picture of the composition of the surface of the Moon as a whole, opening our eyes to large-scale processes rather than regional ones. The first global compositional maps to be constructed showed exactly what would be expected if the Moon had been molten and undergone the separation of materials suggested by the magma ocean hypothesis. Iron is more concentrated in the denser materials that later erupted to the surface to form the maria, so we would expect to see more iron in these regions, and observation now bears this out. While this alone is not proof that a magma ocean did exist, it is one piece of evidence that fits the theory well. Other studies are being carried out to help complete the puzzle of lunar origin and evolution. For example, studies of large impact craters are particularly useful as they excavate material from beneath the surface, giving direct compositional information about subsurface layers and allowing us to probe the vertical structure of the crust (a key factor in models of lunar evolution). Similarly, impact craters on maria may expose several distinct lava flows, and studies of these allow us to infer the volcanic history of an area. On a larger scale, this will help us address the question of the origin, evolution and distribution of lunar volcanism, another unresolved issue in the evolution of the Moon.

In addition to multispectral analyses, topographic data from Clementine is helping to study some of the oldest and most degraded impact basins on the Moon. Using this, scientists have confirmed the existence of the largest known impact basin in the Solar System on the farside of the Moon (Figure 6.2). Called the South Pole–Aitken basin, the structure is an incredible 2500 km in diameter, and may have excavated right the way through the crust to expose material from the lunar mantle. If this is the case, then studying this feature will allow us to look deeper into the lunar interior than ever before, offering the opportunity to study the vertical structure of the crust in its entirety.

Perhaps the most exciting of all Clementine's discoveries stemmed from images of the lunar poles. The Moon has a very low inclination (1.5°) to the plane of Earth's orbit about the Sun, which means that incident solar radiation is almost horizontal at the poles. If there is a mountain high

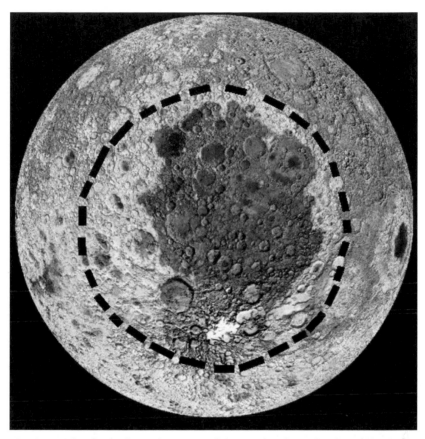

**Figure 6.2.** This figure shows the extent of the South Pole–Aitken basin, the largest known impact crater in the Solar System. It is 2500 km in diameter and covers a large portion of the farside of the Moon. Since impact craters and basins act as probes into the subsurface layers of the Moon, this particular impact allows us to look deeper into the Moon than ever before, perhaps even exposing mantle materials from directly beneath the solid lunar crust. (Image courtesy of the Lunar and Planetary Institute, Houston.)

enough in these regions, it will be bathed in perpetual sunlight. Conversely, if there is an area deep enough (e.g. in the well of a crater) it will remain in permanent shadow, and such areas would be cold enough for water ice to be stable, an incredibly exciting prospect. Clementine images revealed several areas in the south polar region that could lie in perpetual darkness (Figure 6.3), covering an estimated area of 30 000 square

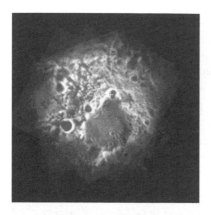 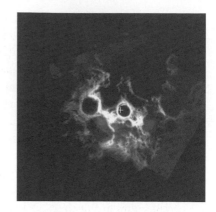

North polar composite          South polar composite

**Figure 6.3.** These images are composites of the North and South Polar regions of the Moon as seen by Clementine over a period of about two months. The southern region has many more shadowed areas than the north, and it is in these permanently shadowed areas that water ice may be found. The Clementine mission detected a signal over the South Pole that may have been due to the presence of water. The Lunar Prospector mission detected large deposits of hydrogen over both poles, and it is suggested that this is locked up in the form of water ice which is mixed in with the lunar soils. If this is the case, then it will certainly be a crucial resource for future lunar explorers. (Image courtesy of the Lunar and Planetary Institute, Houston.)

km. Still more exciting was the news that another experiment on board Clementine had enabled the detection of a signal that *may* have been due to the presence of water ice in this region. This tantalising hypothesis could not be proven with the instrumentation on board Clementine, but fortunately another mission was already in the pipeline that could shed more light on the situation.

Four years on from Clementine, in 1998, NASA launched Lunar Prospector, their first specifically targeted lunar mission since the end of the Apollo program in 1972. Prospector is classed as a *Discovery* mission; designed, built, launched and operated on as low a budget and short a timescale as possible while maintaining an excellent standard of scientific return. It carried just five instruments with focused scientific objectives designed to complement existing datasets, providing the means to answer some of the key questions remaining in lunar science.

Prospector made observations in lunar orbit for well over a year, much longer than any previous mission to the Moon. As with Clementine, observations were global, and with an orbit just 100 km above the lunar surface it is also the closest any orbiting spacecraft has come to the lunar surface for a sustained period. The first results from Prospector were published in a special edition of the journal *Science* in 1998, and are still being analysed in the context of how they can be used to refine and update existing lunar models. Possibly the most important measurement is the mapping of the distribution of elements across the Moon using the gamma-ray spectrometer which could detect elements that make up an estimated 98 per cent of all lunar materials. It is hoped that these data in particular will help us to refine our estimates of the bulk composition of the Moon, still unknown today. Similarly, knowledge of the distribution of trace elements is important in testing theories for lunar evolution, as some of these may have been concentrated in the last melt remaining after the formation of the lunar crust from the magma ocean. The gamma-ray spectrometer on board Prospector was specifically designed to look for the signature of some of these trace elements, and found its distribution to be supportive of this theory. From a more visionary perspective, the global maps of elemental abundance may also be used as a resource for the location and utilisation of construction materials when we start to build lunar bases.

The gravity experiment on board Prospector provided the most accurate gravity maps ever obtained for the Moon, allowing for the detection of irregularities in density that can in turn tell us about the structure of the Moon. As well as this, the fluctuations in density can be used to show us how the pull of the Moon exerts changes on a spacecraft's orbit. This will be very useful for lunar visitors of the future, allowing them to refine their orbital parameters or calculate the amount of fuel required for their journey more precisely. Prospector discovered three new anomalies, and there is a hint of four others on the farside. Perhaps more importantly, coupled with data from the magnetometer, the gravity experiment has provided estimates of the size of the lunar core. It is believed to be between 220 and 450 km in diameter, and will be a critical factor to the theories of both lunar origin and evolution.

Probably the best known and most provocative 'discovery' of Lunar Prospector came with confirmation of Clementine's suggestion that water ice deposits may be present at the South Pole of the Moon; an extra twist came with the discovery of similar deposits at the North Pole (Figure 6.4).

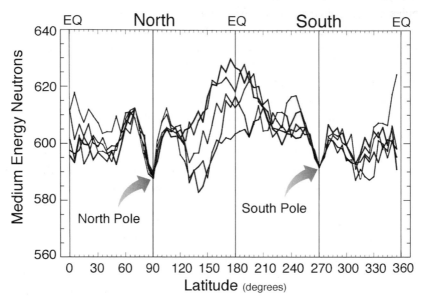

**Figure 6.4.** A graph showing the measurements made by the neutron spectrometer on board Lunar Prospector. A sudden drop is seen in the measured number of 'epithermal neutrons' at both the North and South Poles of the Moon, which indicates the presence of large concentrations of hydrogen which is most likely to be locked up in molecules of water. It is estimated from these measurements that there could be up to three billion tonnes of water ice on the Moon; this is enough to provide around 2000 people with enough water for drinking, washing and preparing food etc. for 500 years. (Image courtesy of NASA Ames Research Center.)

The neutron spectrometer on Prospector is not able to detect water directly, but can locate concentrations of hydrogen, and, although we know hydrogen can be locked up in many molecules, it is believed that the conditions on the Moon make water ice the most likely interpretation. If these results *do* indicate the presence of water ice, then there could be as much as three billion tonnes at each pole, although Prospector detected a greater concentration at the North Pole than at the South. To put this into perspective, it is estimated that each person in London uses 55 000 litres of water per year for drinking, washing, preparing food etc. If we could take just 1 per cent of the Moon's water, it would support 2000 such Londoners for over 500 years. Hence, if direct confirmation of the presence of water at the poles can be obtained, it will have huge implications for the long-term future of lunar exploration. It will certainly make the Moon more appealing for future manned missions, perhaps involving the construction of lunar bases and industrial outposts, and lunar science is a guaranteed beneficiary of all such missions.

After Lunar Prospector completed its primary one-year mission it dropped to an altitude of just 40 km where it remained for a time before moving closer still to just 25–30 km above the surface. Throughout this period, the instruments on board continued to gather data of increasingly high resolution, allowing for the construction of more accurate maps and for refined estimates of the concentration of hydrogen deposits at the poles to be made.

The Clementine and Prospector datasets collectively provide us with the opportunity to unravel some of the Moon's (and the Solar System's) most complex problems. However, there is still a good deal more to learn, and it is important that the recent interest in lunar science is maintained if we are to complete the puzzle. There are more missions to the Moon planned, but it is difficult to look forward more than 10 years in this climate of financial uncertainty. Even so, the following section gives one view of what the future may hold for us, including the known and the unknown; it will be interesting to come back to this chapter in 30 years and see how far we have come.

## 6.3 Visions of the future

The final years of the twentieth century saw us develop a greater understanding of the Moon than we had ever had before. However, as with all science, for every question answered, many more are raised and there is still much work to do if we are to fully understand the origin and evolution of our satellite and place it in context with the rest of the Solar System. Some of the key questions that remain may be addressed by upcoming missions. Within the next few years, Japan intends to launch the Lunar-A mission, designed to fire penetrometers into the lunar surface to monitor seismic activity over its one-year lifetime. Seismic data has not been collected from the Moon since the days of Apollo, and this mission will provide exciting information regarding the structure of the lunar interior and how much the Moon suffers from its equivalent to earthquakes. Europe did not go to the Moon in the twentieth century, but will do so in the first years of the twenty-first. SMART-1 (Small Missions for Advanced Research Technology) is a technology demonstration mission to the Moon funded by the European Space Agency, and the science to be carried out will build on the work completed by Lunar Prospector and Clementine. The only other lunar mission currently planned is another Japanese venture called Selene.

Together, these new missions will certainly fill some gaps in our knowledge of the Moon. However, some of the most fundamental questions cannot be addressed without a major, long-term effort, and will probably require highly sophisticated unmanned craft and/or manned missions. For example, on a global scale, even though the Clementine and Prospector datasets have provided good insights into the crustal composition and topography of the Moon, a *detailed* assessment of the composition of the lunar crust will require the return of a comprehensive set of lunar rock and soil samples to Earth, representative of the Moon as a whole. The current sample collection includes less than half of the types of mare basalts known to be present on the nearside of the Moon, and the farside remains completely unsampled. Another problem with the samples currently in our possession is that they were all found as debris strewn across the lunar surface as a result of the impact process. Consequently, we have no way of ascertaining what their original surroundings would have been, and cannot construct a detailed picture of where they lie in terms of the timeline for crustal formation or volcanic history. With the orbital data now available, sites of interest can be selected which have the best chance to characterise a particular area in composition and/or age, and these should be carefully explored, taking samples of rock from their natural surroundings. A good example of where such analyses would have been useful is the Apollo 15 landing site. The lunar module touched down near a volcanic feature called a sinuous rille, believed to have been created by turbulent lava flows. The rille in this area, called Hadley Rille (Figure 6.5) clearly showed layering of the rocks within the walls. If samples could have been taken from within these layers, we would have been able to determine the relationship between the layers in age and composition, something we have yet to achieve anywhere on the Moon. Knowledge of an accurate timeline for the events during which these and other features were formed would give us a clear insight into the most active periods in the Moon's history, both internally (e.g. volcanic activity) and externally (e.g. impact cratering events).

**Figure 6.5.** The sinuous channel in this image is called Hadley Rille, and was the landing site of the Apollo 15 mission in 1971. The rille is 100 km long, and averages 1.5 km wide and 400 m deep. Lunar rilles are thought to have formed through the action of flowing lava, and the investigations of the Apollo 15 astronauts in parts of Hadley clearly showed layering of ancient volcanic flows, indicating that volcanism on the Moon had taken place over a period of time and was made up of many individual events. *In situ* measurements of materials from layers such as these are vital if we are to gain a clear picture of the history of our Moon and place the samples collected in their correct geological context. (Image courtesy of NASA.)

The method by which an increased sample database could or should be collected is a contentious issue. While samples could be returned by unmanned craft, the limitations inherent to this technique would strongly compromise the overall science return. The sharp contrast in the quantity, quality and diversity of the science returned from the manned Apollo missions compared to that of unmanned sample-return programmes, is a powerful indication of the merits of manned spaceflight. This point is underlined by the three unmanned sample-return missions to the Moon in the 1970s, whose combined return of 300 g of lunar soil amounts to less than 0.1 per cent of the total sample collection currently in our possession. In addition, because the craft were unable to rove around the surface, the samples could not be selectively chosen and had to be taken from directly under the landing site. While it is true that technology has come a long way, and the hugely successful Pathfinder mission to Mars with its Sojourner rover has proven that unmanned missions need not be restricted to the spot where the mother craft landed, it is impossible to question the superior ability of a human being to make on-the-spot decisions, choosing specific samples based on years of training or experience that simply could not be programmed into a rover or unmanned craft. In spite of the advances made in rover technology, humans are also more manoeuvrable over rough terrain and can survey a much greater area. No machine now or in the immediate future has or will have the capability to observe, assess, select and preliminarily analyse rocks and samples to the degree that humans can. The arguments against sending humans into space are primarily economic, political and moral. Unmanned craft are substantially less expensive to send into space and do not require life support systems. Also there is very little room for error in a manned mission, and it is infinitely preferable for a machine to 'die' than a human on a mission to the Moon (or anywhere else). These are strictly non-scientific arguments however, and are best debated elsewhere. From a purely scientific viewpoint, manned spaceflight to the Moon is unquestionably the way forward in the twenty-first century.

Should manned lunar missions resume (and they should), the eventual goal would be to establish a base large enough to house several astronauts for extended periods of time, from weeks to months (analogous to the International Space Station in Earth orbit). From such a base, geological field excursions could take place, with samples analysed on the Moon itself, giving the opportunity for more interesting finds to be collected in a short space of time. Lunar science would not be the only beneficiary of such a base. Biological experiments could be carried out on plants, animals and humans over extended periods in reduced gravity conditions. Astronomy would benefit hugely from placing telescopes on the Moon; it is stable with a negligible atmosphere and has nights lasting approximately

**Figure 6.6.** An artist's impression of a processing plant on the Moon, extracting and utilising lunar resources. In the not-too-distant future, mankind could establish bases on the Moon for scientific, technical, and perhaps even commercial purposes. Currently the largest hurdle is the cost; but while it is hugely expensive to send a manned mission to the Moon, it is the next logical step forward in the exploration of our natural satellite, and is necessary if we are to answer the fundamental questions that remain about its origin and evolution. (Image courtesy of NASA Johnson Space Center.)

14 Earth-days. Shielded from the Earth, the farside of the Moon is the only radio quiet area in the inner Solar System, providing the perfect platform for radio astronomy. There would be many benefits to having a base on the Moon, not least to gain the experience of building and operating it in preparation for future manned missions, perhaps to Mars. Although the work would be difficult, the construction of a lunar base is probably not too far beyond our technological capabilities even now, and we are already gaining experience in constructing objects in space through the International Space Station (ISS) project. A base is perhaps the next 'giant leap' in our exploration of the Moon, but the huge cost of transporting the materials required from Earth is one of the primary factors currently preventing this from

happening. Before we can seriously consider manned exploration of the Moon in terms of bases occupied for periods of weeks or longer, we need cheap, easy and regular access to space. Hopefully this can be achieved in the next decade or two and we can then start to explore the Moon in greater depth and in more ways than ever before. Only then will we be able to piece together the complex puzzle that will provide a comprehensive view of our natural satellite.

## 6.4  Further reading

Spudis, P. D. 1996 *The once and future Moon.* Washington: Smithsonian Institution Press.

Crawford, I. A. 1998 The scientific case for human spaceflight. *Astron. Geophys.* **39**(6), 14–17.

Harland, D. M. 1999 *Exploring the Moon: the Apollo expeditions.* Chichester: Springer-Praxis.

Mendell, W. W. (ed.) 1985 *Lunar bases and space activities of the 21st century.* Houston: Lunar and Planetary Institute.

Light, M. 1999 *Full Moon.* London: Jonathan Cape.

Dunkin, S. K., Heather D. J. 1999 Unveiling the face of the Moon: new views and future prospects. *Phil. Trans. R. Soc. Lond.* A **357**, 3319–3333.

# 7

# The Earth's deep interior

## Lidunka Vočadlo and David Dobson

*Department of Geological Sciences, University College London, Gower Street, London WC1E 6BT, UK (l.vocadlo@ucl.ac.uk, d.dobson@ucl.ac.uk)*

## 7.1 Introduction

The Earth upon which we live is a poorly understood planet, extending some 6400 km to the centre where the conditions of pressure reach over three million times the pressure we experience at the surface, and the temperature may exceed 6000°C. We stand on thin brittle crustal plates moving through geological time over a continuously deforming mantle of slowly convecting hot rock. Living on its surface, we drill into its crust, and are mere observers of remarkable natural phenomena such as the volcanic eruptions that occur close to plate boundaries. The material which we can sample directly in this way has come from only a few hundred kilometres down into the Earth, from inclusions in diamonds which are brought up to the surface by volcanic intrusions. Yet there remains well over 6000 km to go before the centre is reached – 90 per cent of the Earth is effectively inaccessible. The only way in which we can explore the deep interior directly is by observing seismic waves that travel through the Earth, generated after earthquakes. By analysing these waves, seismologists have built up a picture of the gross structure of the Earth's interior (Figure 7.1). This shows the Earth to be layered with significant seismic wave velocity discontinuities at the boundaries between the different layers – the complex crust, the silicate rock mantle (subdivided into upper mantle, transition zone and lower mantle), and the liquid outer core and solid inner core which are made mostly of iron alloyed with light elements such as sulphur and silicon.

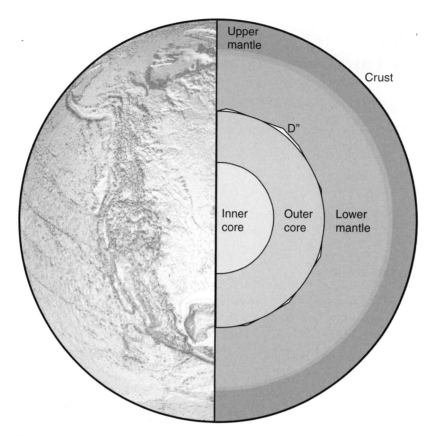

**Figure 7.1.** Schematic view of the Earth's interior depicting the seismically determined layers. The crust and mantle are composed of solid silicate and oxide minerals. The mantle is subdivided by seismic discontinuities into upper mantle (<410 km depth), transition zone (410–670 km) and lower mantle (670–2890 km). The liquid iron outer core (2890–5150 km) contains approximately 10 wt per cent of light alloying elements and the solid iron inner core extends to the Earth's centre at 6371 km depth. The D″ zone at the base of the lower mantle has complex topography ranging from a few tens to hundreds of kilometres thickness, and is thought to be the result of reaction between the metallic core and silicate mantle.

Ideas about the nature of the Earth's interior have developed significantly over recent years. Perhaps the greatest single advance in geological thought during the twentieth century was the development of plate tectonic theory. This elegant model explained the existence and topography of the Earth's oceans and continents, the relative motions of the continents, the compositional variations seen in volcanoes from different tectonic settings, and demonstrated the main mechanism for heat loss in the Earth. It is already clear that many of the large-scale geological processes responsible for the conditions at the surface are driven from the Earth's core. However, many challenges remain in our understanding of Earth processes occurring in the deep interior. For example, we have yet to fully define the composition, convective regime and transport properties of the deep mantle and core. We also do not know how the temperature varies throughout the deep Earth. The nature of the core–mantle boundary region is still a mystery, as are the processes governing convection in the outer core which leads to magnetic field generation.

These problems are all intimately tied up with the behaviour of the Earth's high pressure minerals, and, although we cannot sample deep Earth materials directly, many of the physical properties of these minerals may be determined both in the laboratory and via theoretical computer simulations. The experimental and theoretical techniques that have been used to probe our planet have developed significantly over recent years. Current developments in high-pressure experimental techniques, the rapid advances made by computer modellers, and a cross-disciplinary approach combining these with solid-Earth geophysical methods hold the key to solving the major questions of Earth evolution.

Experimentalists continue to gain access to higher temperatures and pressures allowing laboratory simulation of the physical conditions from the surface of the Earth to the core, shedding light on the physics and chemistry of the Earth's deep interior. Even so, with increasing depth, it becomes increasingly difficult to mimic the extreme conditions of pressure and temperature precisely. An alternative to laboratory experiments is the use of computer simulations which allow us to test which models best match the seismic evidence and experimental data. Computational mineralogy is fast becoming the most effective and quantitatively accurate method for successfully determining mineral structures, properties and processes at the extreme conditions of the Earth's deep interior. Using a variety of simulation techniques, theorists are not only able to provide a microscopic underpinning to existing experimental data, but also provide a sound basis upon which to extrapolate beyond the limitations of current experimental methods. This approach allows us to predict the properties of candidate mantle materials with remarkable accuracy when compared with seismic data and the results of laboratory experiments. With these simulation techniques, we can also try to solve many problems that are out of the reach of experimentation involving simultaneously high pressures and temperatures. For example, we are beginning to understand the nature of iron and iron alloys under the extreme conditions of the core where iron is squeezed to about half its normal volume. We are also starting to provide constraints on the temperature profile of the Earth which, at core depths, is known only to within a few thousand degrees!

With an interdisciplinary approach involving theory, experiment and seismology we will soon be able to determine the nature and evolution of the Earth's deep interior and the influence it has on the surface upon which we live.

## 7.2  Experiments used to probe the Earth's deep interior

Although the Earth's deep interior cannot be readily accessed for study directly, it is possible to produce the high pressure and temperature conditions of the Earth's interior in the laboratory and observe their effects on minerals. The aim of the experimentalist is to achieve these conditions of very high pressures and temperatures with the greatest accuracy and precision, and then probe the microstructure of the mineral samples, looking for clues as to what compositions and processes dominate in the Earth's deep interior. The main challenge, therefore, for experimentalists is to generate the required pressures and temperatures. The two main approaches used to generate ultra-high pressures and temperatures in the laboratory are (i) static compression, whereby a sample is squashed hydrostatically and then heated to very high temperatures either within a furnace or with a laser and (ii) shock compression, whereby a dynamic pressure wave is sent through a sample causing it to rapidly become heated and compressed simultaneously.

### 7.2.1  Static compression

Static compression uses hard anvils to compress a mineral sample, often amplifying the applied pressure by tapering the anvil to reduce the surface area. The simplest and most elegant of the tapered anvil devices is the diamond anvil cell (Figure 7.2). Diamond is transparent to electromagnetic radiation from the far infrared through visible and ultraviolet into the hard X-ray region. Therefore, in addition to direct observation of pressure- and

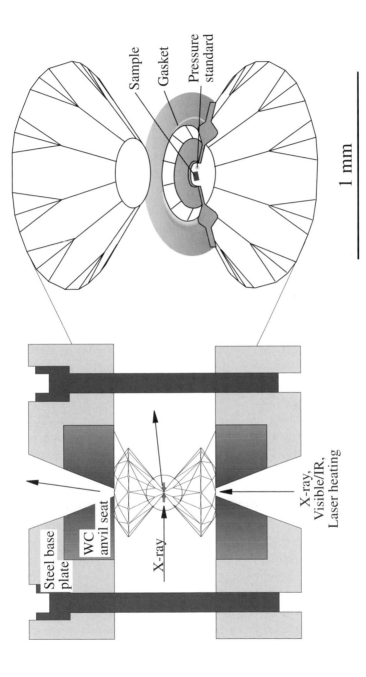

**Figure 7.2.** The diamond anvil cell. Force is applied to the back of the diamonds by tightening the screws connecting the base plates. The diamonds compress the small sample volume between their tips and the metal gasket, shown in the enlargement. The sample and a pressure standard are surrounded by a soft pressure medium and experience nearly perfect hydrostatic pressure. The excellent transparency of diamond to many wavelengths of light allows many spectroscopic and diffraction-based studies to be performed in the diamond cell.

temperature-induced structural changes (phase changes) in minerals, these devices are ideal for a wide range of spectroscopic and diffraction-based techniques applied while the sample is under pressure; such *in situ* measurements are ideal for probing the microstructure of the sample.

In the diamond anvil cell, pressure is applied by way of a small load on the backs of two opposing gem-cut diamonds which surround the sample. The small pressure on the back of the diamonds is amplified many times at their tips, which have a truncated area of $0.2\,mm^2$ or less. The volume at high pressure, typically $10^{-3}\,mm^3$, is compressed uniaxially by the diamonds. Within this volume, a pressure medium (liquid methanol–ethanol mixture for low pressures and temperatures, or high-density gas for high pressures and temperatures) surrounds the sample $(10^{-4}\,mm^3)$ and exerts a hydrostatic pressure. A gasket stops the pressure medium from extruding and seals the sample environment. In this way, pressures in excess of 300 GPa (1 GPa is approximately 10000 times atmospheric pressure) can be obtained in the diamond cell. In order to reproduce the conditions of the Earth's deep interior, the sample must also be heated. In the diamond anvil cell, this can be achieved either by heating indirectly to moderate temperatures ($<1000\,K$) using a resistive furnace placed around the diamonds or directly to high temperatures ($>4000\,K$) by firing a laser beam into the sample.

## 7.2.2 Shock compression

Shock compression is achieved by essentially firing a gun at a target and analysing the deformed sample, which has been subjected to very high pressures and temperatures, after the impact. A high-velocity projectile is fired at the sample and the impact shock wave generates ultra-high pressures and temperatures in the sample for very short times of milliseconds or less. In such shock experiments, the impact generates pressures up to a billion times greater than atmospheric pressure. Figure 7.3 shows a two-stage light-gas gun consisting of two long gun barrels connected end-to-end via a solid disc which is designed to rupture at a critical pressure. The small-bore portion contains the projectile, adjacent to the rupture disc, and opens in to the evacuated sample chamber. An explosive is placed at the far end of the large-bore barrel, behind a light piston. The experiment is initiated by igniting the explosive which accelerates the piston down the large-bore tube and compresses a light gas between the piston and rupture disc. When the disc ruptures the gas expands into the small-bore barrel,

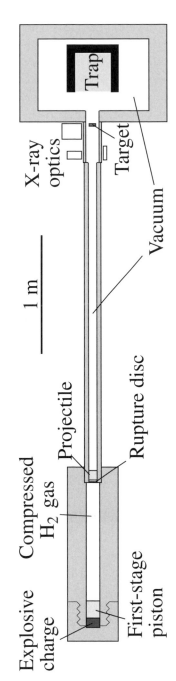

**Figure 7.3.** The double-stage light-gas gun can generate ultra-high pressures and temperatures in the terapascal and kilokelvin range for durations of up to milliseconds. The experiment is initiated by igniting the charge which drives the piston along the first stage (large-bore) barrel. The hydrogen gas is compressed by the piston until the disc connecting the first and second stage barrels ruptures. The expansion of the gas accelerates the projectile along the second stage (small-bore) barrel and into the sample at velocities of several kilometres per second. The resultant shock wave generates ultra-high pressures with adiabatic heating in the sample. Fragments of the sample can be collected in the trap for later analysis. The projectile velocity is measured using mirrors connected to the sample. Shock velocity is measured using two beams of light, which cross its path with a precisely known spacing. Shock velocity is measured using mirrors connected to the sample and a high-speed film.

accelerating the projectile before it. The maximum attainable velocity is determined by the rate at which the gas can expand, hence the use of a light gas, usually hydrogen. The projectile exits the small-bore barrel and impacts the sample causing a shock wave and heating; the sample then disintegrates during pressure release. The extraordinarily high pressures last only for a very short time (milliseconds), and, although it is very difficult to determine precisely the pressures and, particularly, the temperatures existing within this complex sample environment, the ultra-high pressure range achieved renders these experiments invaluable for studying compression and structural changes at the conditions of the Earth's deep lower mantle and core.

### 7.2.3  The experiments of the future

The development over recent decades of *in situ* techniques to measure a range of physical properties at high experimental pressures and temperatures have opened a whole new vista for the deep-Earth sciences. Optical and gamma-ray spectroscopy in the diamond cell provide powerful probes of the electronic and nuclear environment of samples, while X-ray diffraction in high-pressure devices has been the work-horse of high-pressure crystallographers determining crystal structures and compressional behaviour for several decades. Electrical conductivity measurements with electrodes passing through the gaskets or embedded within high-pressure devices allow *in situ* measurements of the mobilities of various species within minerals and provide important constraints for interpreting geomagnetic field measurements in terms of electrical, thermal and chemical structure within the deep Earth.

It is, however, the advent of synchrotron radiation sources with dedicated high-pressure beamlines which will most revolutionise high-pressure experiments in the coming decades. A synchrotron provides high-energy, high-brightness radiation which is required for frontier experimentation allowing *in situ* high pressures and high temperatures which enable us to probe the short-range structural and electronic interactions between atoms in minerals.

A synchrotron works by accelerating clusters of charged particles around a racetrack-shaped booster synchrotron several hundreds of metres long to speeds close to the speed of light (Figure 7.4). The particles are then injected into a storage ring of over one kilometre circumference. The particle beam orbits the storage ring several hundred thousand times per

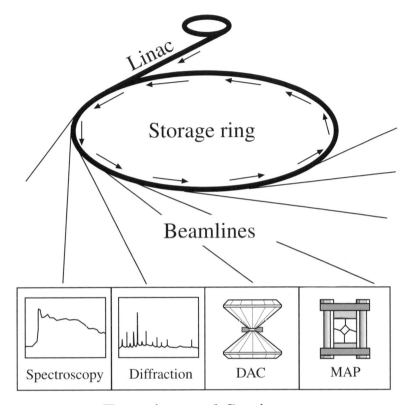

Spectroscopy | Diffraction | DAC | MAP

## Experimental Stations

**Figure 7.4.** Schematic of a synchrotron facility. Bunches of electrons or positrons are accelerated to near light speed and injected into the storage ring using the linear accelerator (linac). These bunches of charged particles emit broad spectrum X-rays at bending magnets, or more highly conditioned X-rays (single wavelength, polarised) at insertion devices on the straight segments of the storage ring, between the bending magnets. The emitted X-rays are collimated to beamlines with very low angular divergence and continue to the experimental stations. The very high intensity, high energy and conditioning of synchroton radiation makes it ideally suited to a range of spectroscopic and diffraction-based techniques. In addition, the intensity penetrates the pressure medium in high-pressure experiments (diamond anvil cell (DAC) or large volume multi-anvil press (MAP)) and allows *in situ* measurements at high pressure and temperature to be performed.

second, steered and focused by powerful electromagnets. As the beam decelerates it emits high-intensity radiation in a spectrum from infrared through to gamma rays – this high-energy radiation is synchrotron radiation. The synchrotron radiation is emitted in a direction tangential to the arc of the particle beam towards optical devices which select the desired wavelength and then pass it on down the beamline to various experimental stations and to the sample under investigation.

Synchrotron-based X-ray diffraction in diamond anvil cells allows rapid mapping of pressure and temperature stability and structural refinement of minerals. The addition of diffraction-based pressure standards to the sample allows very accurate determinations of phase boundaries, which is particularly important for comparing pressure-induced mineralogical phase changes with seismic discontinuities. Simultaneous measurements of ultrasonic sound-wave velocity with diffraction will increase our precision in these high-pressure/high-temperature studies by orders of magnitude and will allow quantitative measurements of the high-order terms which are vital for mapping mineral compression across the pressure range encountered within the Earth. Additionally, the speed at which spectra can be collected from synchrotron sources facilitates studies of the kinetics of phase transformations at high pressures and temperatures, and it is also possible to tune X-ray sources for X-ray absorption spectroscopy, thereby allowing the probing of the immediate atomic environment.

Despite the rapid advances being made, there are limits to experimental investigations; for example, it is generally very difficult to establish the detailed microscopic atomistic and electronic mechanisms involved in thermoelastic and dynamic processes simply from experimental study. Furthermore, even though diamond anvil cell techniques are advancing dramatically, the experimental study of minerals under core conditions is still a major challenge. As a result, therefore, in order to complement existing experimental studies and to extend the range of pressure and temperature over which we can model the Earth, computational mineralogy has, in the past decade, become an established and growing discipline.

## 7.3  Computation methods used to probe the Earth's deep interior

Computational methods employ the fundamental principle that the most stable atomic environment is the one with the lowest energy. High-pressure/high-temperature properties of minerals may be obtained through knowledge of the energy of the system. Therefore it is the goal of computer modellers to calculate the energetics of the system of interest and thereby determine what environment is energetically the most favourable at a given pressure and temperature. Once the energy of a system has been established, it is possible to calculate a variety of properties which determine the evolution and characteristics of many geophysical processes. The simulation techniques used in order to model minerals and the processes occurring within the Earth's deep interior generally fall into two categories: *atomistic simulations*, using potential models to describe the interactions between atoms; and *quantum mechanical simulations*, calculating electronic structure from first principles. In the first case, the energy of the system is written in terms of an atomistic potential function (which describes the interaction between the atoms), the parameters for which may be determined either experimentally or from first principles; in the second case, the energy of the system is written in terms of electronic interactions, where appropriate substitutions are made in order to solve the Schrödinger equation for the system. In the latter case, no empirical input parameters are required and an exact solution is found. The chosen technique depends upon the system and process simulated, and the level of approximation required. By setting up on a computer an artificial simulation box of particles, whose energy is governed by either an atomistic or quantum mechanical potential, calculations may be performed in order to predict how these systems respond to high pressures and temperatures.

### 7.3.1  Simulations using atomistic potential models

The interatomic potential gives the energy of interaction between the atoms or ions within a system as a function of their separation and orientation (Figure 7.5). When no net forces are acting on the constituent atoms, the sum of the attractive and repulsive potential energies between each pair of atoms in a crystalline solid at zero kelvin is termed the static lattice energy. This is made up of contributions from the long-range Coulombic attraction between positive and negative ions and a repulsive term due to the diffuse nature of the electron clouds surrounding the nucleus.

The optimum potential function that describes interactions between atoms in a given system is obtained either empirically or from first principles. The system is set up in the computer by creating an artificial box of atoms of the particular mineral of interest. Zero kelvin structural data may

**Energy**

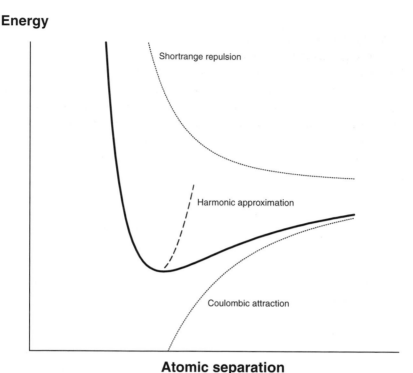

**Atomic separation**

**Figure 7.5.** Energy as a function of atomic separation with both the attractive and repulsive components; at higher temperatures, the atoms vibrate anharmonically, away from the harmonic approximation.

readily be obtained from the potential function. However, in order to successfully model deep-Earth mineral phases, it is desirable to simulate conditions of extreme pressure and temperature. Modelling pressure is fairly straightforward, being derived from the forces between the atoms which are obtained from the gradient of the potential function. However, modelling temperature (and the associated kinetic pressure due to atomic vibrations) requires more sophisticated techniques. The techniques that have been developed in order to achieve this include (i) *lattice dynamics*, where the system is described in terms of atomic lattice vibrations (phonons) each with a wavelength and a frequency, and (ii) *molecular dynamics*, where the atoms are given initial positions and velocities which are allowed to evolve over a period of time via solutions to Newton's equations of motion.

## Lattice dynamics

The lattice dynamics method describes the system in terms of a simulation box containing vibrating particles whose frequencies vary with volume. The motions of the individual particles are treated collectively as lattice vibrations or phonons (Figure 7.6a). The phonon frequencies may be calculated from analysis of the forces between the atoms, and many thermodynamic properties, such as free energies, heat capacities, etc., may be calculated at some desired temperature using standard statistical mechanical relations, which are direct functions only of these vibrational frequencies.

A limitation of the lattice dynamics method is that it treats the atoms as harmonic oscillators (i.e. the atoms vibrate symmetrically about their equilibrium position), whereas at very high temperature the oscillations become anharmonic as their displacements from equilibrium become asymmetric (as seen in Figure 7.5), so another treatment is required to take this into account.

## Molecular dynamics

Molecular dynamics is routinely used for medium- to high-temperature simulations of minerals where atomic vibrations become more anharmonic as the atoms vibrate further away from their equilibrium positions. The interactions between the atoms within the system are described in terms of the potential model discussed earlier, but instead of treating the atomic motions in terms of lattice vibrations, each ion is treated individually. In principle, Newton's equations of motion are solved for a number of particles within a simulation box to generate time-dependent trajectories and the associated positions and velocities which evolve over time (Figure 7.6b). Here, the kinetic energy, and therefore temperature, is obtained directly from the velocities of the individual particles. With this explicit particle motion, the anharmonicity is implicitly accounted for at high temperatures. Molecular dynamics methods may also be used to simulate liquids, and are therefore the preferred method when simulating liquid iron alloys at the conditions of the outer core.

The shortcomings of atomistic simulations lie in their dependence upon the accuracy of the chosen potential model, which, in turn, is limited by the ability of such a potential to describe anything other than atomic systems; no information is obtained about the sub-atomic environment.

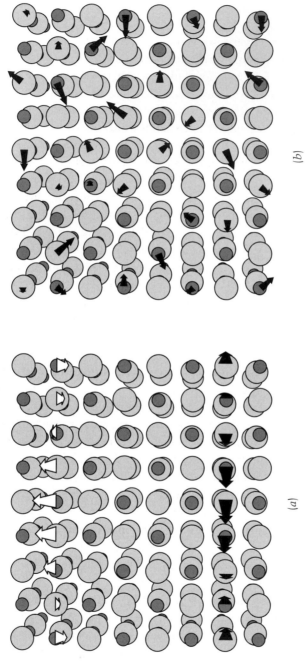

$(a)$

$(b)$

**Figure 7.6.** Schematic simulation boxes for $(a)$ lattice dynamics and $(b)$ molecular dynamics. $(a)$ In a lattice dynamics calculation, at low simulated temperatures, the motions of the individual particles are treated collectively as a vibrational wave (phonon) in the crystal. $(b)$ Molecular dynamics simulations, which take into account the explicit motion of individual particles in the simulation box, are more appropriate for high-temperature simulations since they allow for high-temperature anharmonicity.

Ideally, we would like to model deep-Earth mineral phases without any recourse to empirical parameterisation. This *ab initio* approach is becoming increasingly routine for the simulation of simple systems and simple processes, and is currently being developed to successfully model complex structures with ionic, covalent and metallic bonding under extreme conditions of pressure and temperature.

### 7.3.2 *Ab initio* quantum mechanical simulations

Quantum mechanical simulations are based on the description of the electrons within a system in terms of a quantum mechanical wave function, $\psi(\mathbf{r})$. Although not directly observable itself, the square of this wave function, $|\psi(\mathbf{r})|^2$ gives the probability of finding an electron at any point $\mathbf{r}$. The energy and dynamics of a single electron are governed by the Schrödinger equation. Ideally we would like to solve the Schrödinger equation for our system of interest. However, in minerals we do not have single particles in free space, but instead many bound electrons which interact with both the ionic nuclei and each other. This poses a serious problem as the wave function for a system containing more than one electron is not just a product of one-electron wave functions, but a very complex many-electron wave function.

In principle, once the wave function for a system has been determined, energy minimisation techniques may then be applied in order to obtain the equilibrium structure for the system under consideration. Unfortunately, the complexity of the many-electron wave function, $\Psi$, means that this type of problem cannot readily be solved. However, there are a number of approximations which may be made to simplify the calculation, whereby good predictions of the structural and electronic properties of materials can be obtained by solving self-consistently the one-electron Schrödinger equation for the system, and then summing these individual contributions over all the electrons in the system. Such approximations reduce the many-body Schrödinger equation to that for one electron surrounded by an *effective* potential associated with the interactions of the surrounding crystal. However, the electron is not in an average field because there is *correlation* between the electrons, i.e the electrons interact with each other, and also electronic spin (the *exchange*), governed by Pauli's exclusion principle, both of which serve to reduce the energy of the system. Therefore, an approximation must made for this electronic exchange and correlation.

Once the energy of the system has been calculated, the interatomic forces may then be obtained, which, when combined with classical lattice dynamics or molecular dynamics techniques mentioned earlier, may be used to calculate thermoelastic and related dynamic properties such as free energy. Using these techniques, it is possible, in principle, to simulate any desired mineral at pressures and temperatures existing throughout the entire Earth. Such calculations are time consuming and computer intensive, but with increasingly powerful supercomputers, these calculations are being performed on ever bigger and more complex systems.

## 7.4 Probing the Earth's deep interior

Using the experimental and theoretical techniques outlined above, it has been possible to probe the Earth's deep interior with increasing accuracy and precision. In this way we are working towards a fuller understanding of the Earth's deep mantle and core, and the reaction between them.

### 7.4.1 The Earth's mantle

The Earth's mantle extends from a depth of a few tens of kilometres to about 2890 km. The minerals existing and the processes occurring in the Earth's mantle are determined by the thermodynamics and energetics of the deep interior. Heat, generated within the Earth by crystallisation of the liquid outer core and by radioactive decay, is transported to the surface mainly by slow convection in the mantle which, over geological timescales, flows plastically. Oceanic crust is constantly being generated at upwelling regions of the mantle convection cells and being destroyed at subduction zones, where it is dragged deep into the mantle on downwelling convective limbs (Figure 7.7). One of the major problems associated with the Earth's mantle is understanding what happens to these subducting slabs. Where do they go? There is considerable debate over whether the subducted oceanic crust continues down to the core–mantle boundary or ponds on top of the denser lower mantle. A mid-mantle reservoir of subducted material would indicate that convection within the mantle occurs in separate upper and lower mantle cells, with little mass transfer across the boundary, and separate compositions in the upper and lower mantle cells. Furthermore, heat would be transferred across the boundary layer by conduction, resulting in a very high thermal gradient of 1000 K or more. Conversely, if material is subducted to the core–mantle boundary, then the mantle convects as a single cell with good chemical mixing between upper and lower mantle.

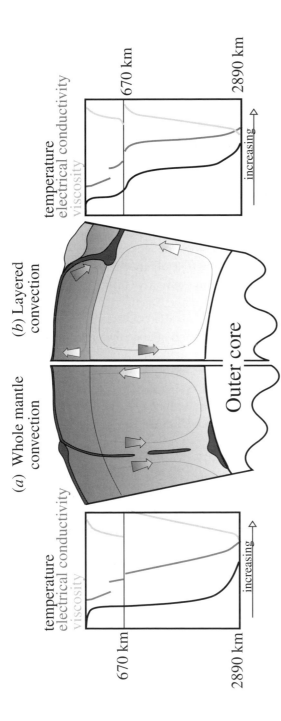

**Figure 7.7.** Schematic of (*a*) whole-mantle and (*b*) layered convection; dashed lines represent flow direction. Inferred temperature (black line), electrical conductivity (grey line) and viscosity (pale line) are plotted against depth for both scenarios. (*a*) In whole-mantle convection, the 670 km seismic discontinuity is due solely to phase transformations to high-pressure polymorphs. There is no mid-mantle thermal discontinuity, with an adiabatic thermal gradient between 200 and 2000 km. Large increases in viscosity and electrical conductivity at 670 (and 410) km depth are postulated due to olivine–spinel and spinel–perovskite + magnesiowüstite phase changes. (*b*) Layered-convection models imply that there is little mass transfer across 670 km, with the upper and lower mantle convecting as separate systems. This results in the upper and lower mantle having separate compositions and a large superadiabatic thermal gradient in the mid-mantle where heat is transferred from the lower mantle to the upper mantle by conduction. Increased lower mantle temperatures, relative to the whole-mantle model, result in increased electrical conductivity and reduced viscosity in the lower mantle.

Seismometers placed over much of the Earth's continental surface allow the measurement of the velocities of compressional (P) and shear (S) waves that travel throughout the Earth after an earthquake. Multi-array seismic studies clearly demonstrate that at 410 and 670 km depth within the Earth's mantle there are significant discontinuities in density, P-wave and S-wave velocity (Figure 7.8). These discontinuities may be produced by a change in composition, or by a change in mineralogy to denser high-pressure phases. Early high-pressure phase relation studies showed phase changes with increasing pressure (and therefore depth) in the dominant magnesium silicate system $Mg_2SiO_4$ from olivine to wadsleyite, ringwoodite and, finally, to $(Fe,Mg)SiO_3$ perovskite plus $(Fe,Mg)O$ magnesiowüstite. Experiments show that the pressures of the transitions olivine–wadsleyite and ringwoodite–perovskite plus magnesiowüstite correspond with the 410 and 670 km depths. Clearly the mantle is seismically and structurally layered, but what about convective layering?

The fate of subducting slabs is highly dependent on whether the mantle is undergoing layered or whole-mantle convection. Recent fluid dynamical simulations of the mantle, which include the phase changes at 410 and 670 km, suggest that the convection within the Earth's mantle periodically switches between stable modes of two-layer convection to convective overturn during which there is significant mass and heat transfer across 670 km. These geophysical simulations are sensitive to estimates of mineral properties, in particular their viscosity and material strength. For example, the ability of a subducting slab to penetrate the density barrier at 670 km depends on the relative strength of olivine and perovskite. We have a qualitative understanding of the strength of mantle minerals; we know that heat weakens minerals and we also know that the high-pressure perovskite phase is significantly stronger than upper mantle phases. However, recent results from diffusion experiments suggest, counter-intuitively, that wadsleyite in the transition zone, between 410 and 670 km depth, may be weaker than either olivine or perovskite. So the effect this will have on the penetrative abilities of subducting slabs into and out of the transition zone is unclear.

Computer simulations can provide constraints on the convective regime in the mantle. Improvements in simulation speed and accuracy allow accurate calculation of the pressure–volume–temperature regime in perovskite and magnesiowüstite. However, comparing this with the density profile derived from seismic studies gives inconclusive results

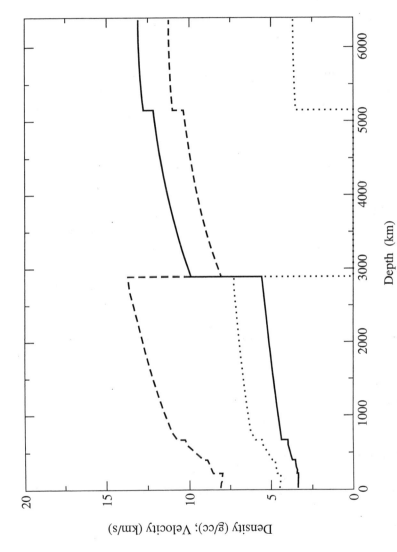

**Figure 7.8.** The variation of density (—), P-wave velocity (- - -) and S-wave velocity (· · ·) with depth, as determined by analysis of multi-array seismic data. Significant discontinuities occur at major boundaries: upper mantle/transition zone at 410 km; transition zone/lower mantle at 670 km; lower mantle/outer core at 2890 km; outer core/inner core at 5150 km.

which show that (i) there is a significant thermal gradient, of 1000 K or more, towards the base of the lower mantle and (ii) a lower and upper mantle of the same composition produce geotherms consistent with whole-mantle convection, whereas a lower mantle significantly enriched in silica is consistent with a 1000 K discontinuity at 670 km and therefore with layered convection.

It is clear that a thorough understanding of the physical properties of mantle minerals, and accurate composition and temperature profiles of the mantle are essential in order to determine the resting place of subducting slabs.

## 7.4.2 The core–mantle boundary

The changes in chemical and physical environment between the Earth's core and mantle are as great as any in the Earth. Density and seismic velocity changes across the core–mantle boundary are more than one order of magnitude larger than across the 410 and 670 km seismic discontinuities combined. The change from solid-silicate lower mantle to liquid-metal outer core entails a change in physical state and chemistry as profound as the atmosphere–geosphere boundary. As well as the major radial variations, the core–mantle boundary shows strong lateral heterogeneities. The core–mantle boundary is intimately connected to evolution of the Earth through the heat transfer and magnetic coupling between outer core and mantle.

Seismic studies of the core–mantle boundary suggest that the base of the lower mantle is highly complex, with zones of seismic anisotropy, seismically slow zones and seismic discontinuities. These regions have a complex topography which may be controlled by mantle convection cells. Seismic anisotropy can be explained by preferential alignment of minerals equilibrating with local stress fields. The seismically slow zone in the bottom 150–450 km of the mantle has been interpreted as being due to partial melting. Seismic discontinuities require rapid variations in density which may be due to (i) infiltration of outer core material into the lower mantle and subsequent reactions between these iron alloys and the lower mantle, (ii) ponding of subducted oceanic lithosphere, or (iii) partial melting of the lower mantle.

In spite of its complexity and importance, the core–mantle boundary is very poorly understood. The large density contrast between core and mantle acts as a barrier to significant mass transport between core and mantle, resulting in a large thermal gradient at the base of the lower mantle. The temperature of the core–mantle boundary is very poorly constrained. In principle, we can anchor the pressure and temperature at the solid–liquid phase change at the inner core boundary and also the spinel to perovskite + magnesiowüstite solid–solid phase transition at 670 km, however, temperature estimates at the core–mantle boundary rely on either extrapolations upwards from the inner core boundary or downwards from 670 km. A large temperature increase near the core–mantle boundary would be gravitationally unstable unless the density of the lower mantle material increased in this region.

Another important result of the core–mantle boundary mass transport barrier is that the outer core and lower mantle may well be significantly out of chemical equilibrium. Chemical reactions between core and mantle materials have been studied in laser-heated diamond cells and have been shown to involve solutions of iron and oxygen into the liquid metal leaving solid FeSi and iron-poor silicate residues. In addition, there may be loss of some of the core alloying elements to the mantle as iron is extracted from the outer core by crystallisation causing enrichment of the alloying light elements. The resulting mixture of iron alloys and iron-depleted mantle silicates at the base of the lower mantle might explain the anomalous seismic behaviour observed in this region of the Earth.

Is the core–mantle boundary where the subducting slabs end up? Or is it just a core–mantle reaction zone? There is no doubt that the core–mantle boundary is an extremely complex and puzzling region of the Earth, and one that holds the key to many of the problems associated with our understanding of the Earth's deep interior.

### 7.4.3 The Earth's core

The Earth's core occupies approximately half of the planet by radius, yet the conditions of pressure (>135 GPa) and temperature (c. 3000–7000 K) make it very difficult to study experimentally. The core is predominantly made up of iron with some lighter alloying elements. In order to understand the core we would ideally like to have information on all multiphase systems containing iron and candidate lighter elements at core conditions. However, even the high-pressure/temperature behaviour of *pure* iron presents major problems which must be resolved before the more complex phases can be properly understood. Therefore, the physical properties of iron and its alloys are of considerable interest to Earth scientists.

## Iron in the inner core

The inner core is crystallising out of the liquid outer core and knowledge of the melting curve for iron at these very high pressures and temperatures would place considerable constraints on the temperature at the inner-core/outer-core boundary. The temperature of crystallisation of the outer core is dependent on the structure of the sold iron phase in the inner core, which itself is a subject of rich debate. Direct experimentation on iron at core pressures (up to 360 GPa) and temperatures (up to 7000 K) has yet to be achieved; even at experimentally accessible conditions (pressure below 200 GPa, temperature below 3–4000 K) there are major conflicts between the results of different groups. In particular, the possibility of solid–solid phase transitions has been observed above c. 35 GPa and 1500 K and above 200 GPa and 4000 K through *in situ* X-ray diffraction experiments; however, the structures of these phases are unknown. This is an ideal example of where calculations can help, since there are no limits to the pressures and temperatures at which the simulations can be carried out, and we can therefore perform first principles calculations on all the suggested candidate phases of iron at core conditions. These simulations show that iron undergoes no phase changes whatsoever throughout the pressure and temperature regime of the inner core, remaining in the hexagonal structure which has been observed experimentally above c. 10 GPa. Moreover, calculations show that this phase of iron melts at 6300 K at inner-core boundary pressures; lighter alloying elements present in the outer core lower this melting temperature by a few hundred degrees, implying that the true temperature at the inner-core boundary is probably 5500–5800 K.

## Iron alloys in the core

The composition of the outer core determines both the temperature at the inner-core/outer-core boundary, where iron is crystallising out of the liquid outer core, and the convective behaviour of the outer core, which is responsible for the Earth's magnetic field. From seismological evidence, the inferred density of the Earth's outer core is too low by some 10 per cent for it to be composed purely of iron. Therefore, a number of lighter alloying elements have been suggested to exist in the outer core, although the exact nature and composition of the alloy is uncertain. Candidates include oxygen, sulphur, silicon, carbon and hydrogen, although it is likely that any, a combination, or all of these could account for the observed density deficit in the outer core. Knowledge of the thermoelastic properties of these iron alloys fundamentally underpins models of planetary formation and critically controls evolution, yet data at even modest pressures and temperatures are scarce both theoretically and experimentally. In order to determine the light element in the core, a number of experimental and simulation studies are currently being undertaken worldwide; however, different laboratories often obtain conflicting results. One of the challenges facing us in the future is the reconciliation of computational and experimental studies of iron alloy systems at extreme pressures and temperatures so that we may finally determine the exact composition of the Earth's core.

### Inner-core anisotropy

The inner core is known to be anisotropic, i.e. the physical properties of the core are dependent upon its orientation. This has been observed from the presence of seismic travel-time anomalies in the core region, but the origin of this anisotropy is unknown. It may be due to small heterogeneities within the core, or intrinsically due to the possible anisotropic structure of the crystalline core material. The latter explanation is often favoured because it is thought that pure iron exhibits elastic anisotropy at core conditions. However, the inner core is unlikely to be one big crystal of iron, but is, instead, a collection of randomly oriented crystals. Such a polycrystalline aggregate of iron will have less anisotropy than a pure crystal, so it is important to know the magnitude of elastic anisotropy in iron in order to constrain the preferred alignment in the inner core. *Ab initio* calculations on iron at inner-core conditions have shown that the P-wave anisotropy of the phase with hexagonal symmetry is identical in symmetry and magnitude to that of the bulk inner core. Although these calculations are a significant step forward in our understanding of inner-core anisotropy, recent seismic studies have suggested that inner-core anisotropy may be an artifact arising from deep-mantle seismological structure, or that the inner core itself has some transitional structure between an isotropic upper inner core and an anisotropic lower inner core. Could the inner core have layers of different structural phases, similar to those already observed in the mantle? Firm evidence of this would require us to radically rethink our ideas of what is actually going on in the very centre of our planet.

## 7.5 Conclusions and outlook

We have reviewed some of the recent exciting developments towards our total understanding of the Earth's deep interior. However, there remain many unsolved problems that need to be resolved if we are to fully understand our planet. For example, although the upper mantle and transition zone are relatively well constrained in composition, geotherm and physiochemical environment, the water content of the transition zone, which could critically affect the melting and transport properties of upper-mantle minerals, is very poorly constrained. As we extend further into the Earth, compositional profiles become less clear; seismic tomography shows compositional heterogeneities in the lower mantle but the timescale of their stability is not known and is dependent on the convective nature of the lower mantle. This, in turn, is dependent on the viscosity of the mantle, but the viscosity of the lower mantle is poorly constrained with estimates ranging from $10^{16}$ to $10^{21}$ Pa s at the top of the lower mantle. Then there are the complexities of the origin of the D″ zone, a region of compositional, geothermal and seismic heterogeneity just above the core–mantle boundary. Is this a reaction zone, or a ponding zone for subducting slabs? We are even less sure of the Earth's core; the composition and geotherm at core depths are highly speculative. Core temperatures are uncertain to within a few thousand kelvin and the exact nature of the 6–10 per cent light elements is unknown. The temperature at which the inner core is crystallising out of the outer core will determine the heat flux into the outer core which, in turn, will determine the convective regime in the outer core that is responsible for the geomagnetic field. The answers to these, and other questions, are not far away.

It is unlikely that we shall ever be able to directly sample the Earth's deep interior; the future of such high-pressure/high-temperature research will depend on the continuing developments in the laboratory that will allow experimentalists to access increasingly higher pressures and temperatures, in conjunction with the rapid advances of supercomputer power that will enable increasingly complex calculations to be performed. The future will see laboratories with microanalytical *in situ* techniques that will not only be able to reach the simultaneously high pressures and temperatures of the Earth's core, but also the extreme conditions to be found in other planets within our Solar System. Advances in the computing industry will enable scientists to have desk-top multi-processor supercomputers so that *ab initio* molecular dynamics simulations on increasingly complex structures will become routine, modelling the properties and processes of planetary materials that critically underpin the formation and evolution of our Solar System.

## 7.6  Further reading

Allen, M. P. & Tildesley, D. J. 1987 *Computer simulation of liquids*. Oxford: Clarendon Press.

Cochran, W. 1973 *The dynamics of atoms in crystals*. Edward Arnold Publishers.

Manghani, M. H. and Syono, Y. (eds.) 1987 *High pressure research in mineral physics*. Tokyo: Terrapub.

Manghani, M. H. & Yagi, T. (eds.) 1998 *Properties of earth and planetary materials at high pressure and temperature*. Washington DC: AGU.

Gurnis, M., Wysession, M. E., Knittle, E. & Buffett, B. A. (eds.) 1998 *The core–mantle boundary region*. Washington DC: AGU.

Vočadlo, L. & Dobson, D. 1999 The Earth's deep interior: advances in theory and experiment. *Phil. Trans. R. Soc. Lond.* A **357**, 3335–3357.

# 8

# Three-dimensional imaging of a dynamic Earth

## Lidia Lonergan[1] and Nicky White[2]

[1]TH Huxley School of Environment, Earth Sciences & Engineering, Imperial College, Prince Consort Road, London SW7 2BP, UK (l.lonergan@ic.ac.uk)
[2]Bullard Laboratories, Department of Earth Sciences, Cambridge University, Madingley Road, Madingley Rise, Cambridge CB3 0EZ, UK (nwhite@esc.cam.ac.uk)

## 8.1 Introduction

Seismic imaging is the most important tool used to investigate the solid Earth beneath our feet. Over the last 20 years, a major advance has been the rapid development and application of three-dimensional (3D) seismic reflection technology. Routinely used by the hydrocarbon industry to aid exploration for, and extraction of, oil and gas, this 3D imaging technique is now ripe for exploitation on a global scale. Seismic reflection surveying uses acoustic or sound energy which is easily transmitted through solid rock. Where rock properties change at depth, some of this energy is reflected back towards the surface, just like an echo, and recorded. Since the 1960s, many important scientific breakthroughs have been made using two-dimensional (2D) seismic imagery. More recently, 3D seismic surveying has become cheaper and coverage has rapidly increased. A typical 3D survey generates about 300 billion bytes of information which, after sophisticated signal processing, yields a cube-shaped image of the subsurface. With this unique probing ability, we can map the 3D subsurface architecture of continental margins where repositories of sedimentary rock contain an important record of how our planet has behaved over millions of years. We can also image the detailed pattern of deformation within these rocks.

In this chapter, we examine the structure of sedimentary basins and show how recent technological developments in seismic imaging have generated spectacular images of what basins look like under the Earth's surface on a variety of scales. Within sedimentary basins, 3D images reveal a range of phenomena: large-scale tilting; folding and faulting of rock layers; and the details of sedimentary strata infilling the basin. Other time-dependent processes such as the movement of molten rocks, oil, and water through the pores of rocks can be monitored by repeated 3D surveying. What makes the 3D seismic technique so powerful is that it is not just a static image of the Earth today: it contains an indirect record of the fourth dimension – geological time.

## 8.2  The convecting Earth

Earth scientists have shown that the horizontal movement of rigid plates at the Earth's surface is driven by slow convection currents deep within the mantle. Rigid plates are fundamental to the theory of plate tectonics which was first proposed in the 1960s. However, the rigidity of these lithospheric plates, which are 120 kilometres thick, prevents us from observing the detailed pattern of convection currents underneath.

One way of learning more about processes deep inside the deep Earth is to exploit the fact that the upward movement of hotter, and downward movement of cooler, mantle material cause subtle vertical motions of the Earth's surface. Thus, erosion of the land and redistribution of sediment by rivers indirectly record regional uplift and sinking. Since uplift is eventually removed by erosion, the most complete record of this complex process is preserved in large depressions called sedimentary basins.

Analysis of amalgamated 3D seismic datasets collected from sedimentary basins worldwide could help to quantify the spatial and temporal variation of vertical motions of the Earth's surface and, by inference, of mantle convection. We suggest that seismic imaging is the key to unravelling elusive yet fundamental processes that keep our convecting planet alive.

## 8.3  What is a seismic image?

The seismic technique uses sound waves which travel through the Earth and are partially reflected at interfaces between rock layers. The first seismic experiment, using a man-made energy source, was carried out by an Irish engineer called Robert Mallet in October 1849 on Killiney Beach near Dublin. His energy sources were 25 pound charges of gunpowder buried at 6 feet depth and the resulting seismic waves were detected using his home-grown 'seismoscope'. The physics which underpins seismic imaging is based upon classical wave theory and was largely developed by the turn of the (last!) century, but the recording instruments were inadequate. Significant advances in seismic technology were made after the Second World War when simple two-dimensional experiments were used to make profiles through the Earth's crust. However, the images produced were of poor quality and frequently noisy. The next significant advance was a method for adding together the recordings to improve the signal, which was patented by scientists from Massachusetts Institute of Technology (MIT) in 1956. This important technique is the basis for all modern seismic reflection experiments and was rapidly adopted by the petroleum industry leading to further significant technical advances. Over the last 20 years, three-dimensional seismic recording has been developed and it is now the routine approach used by the petroleum industry.

The cheapest and best quality seismic data are collected at sea (Figure 8.1). Sound waves are generated by a source suspended in water depths of 5–10 metres off the ship's stern. Most sources consist of several arrays of small airguns which are fired at intervals of 10–20 seconds. A vibrating bubble of compressed air is generated and the resulting shock wave travels through the seabed into the Earth's crust beneath. The energy from airguns is modest – equivalent to about 140 000 kilojoules (or 170 snack bars!) and only a tiny proportion of this energy (5 snack bar's worth) is converted into sound waves.

Most of this energy is transmitted straight through rocks, but a proportion is reflected at interfaces between different rock types where the rock properties, such as density and the speed of sound, change abruptly. Reflections travel back up through the water layer where they are recorded by instruments called hydrophones, located at intervals on a long cable or streamer, towed in a straight line behind the ship. In 2D experiments, a single streamer, which can be anything from 3 to 12 kilometres long, is used. In 3D experiments, a series of parallel traverses are required to build up the 3D image. It is more efficient to collect swathes of data by towing 6–12 streamers behind the ship. Keeping a large number of streamers straight and parallel to each other, particularly in rough seas, is difficult and ships collecting 3D data tow streamers that are 6–8 kilometres long.

Signals returning from rocks beneath the seabed are weak compared with the noise generated by waves and other vessels. This problem is

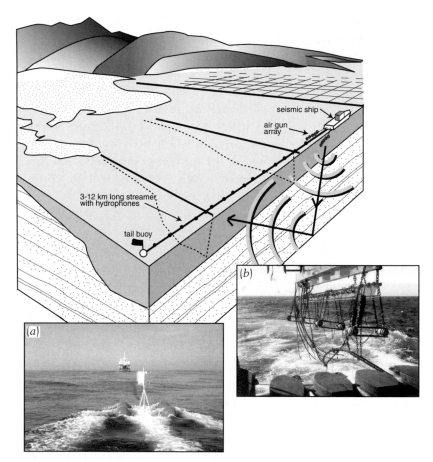

**Figure 8.1.** The essential elements of a seismic reflection experiment. Airguns, which generate sound waves, are towed behind a ship and are suspended 5–10 metres below the sea surface. An oil-filled cable, or streamer, with instruments for recording the returning sound waves (hydrophones) is towed in a straight line and maintained at 10–20 metres depth behind the ship. The grid of dashed lines represents a complete 2D survey. The ship fires its airguns every 10–20 seconds. Data are continuously recorded by the hydrophones and stored in the ship's computers. In a 3D experiment, swathes of data are collected by towing 6–12 streamers. Photographs: (a) View of the ship's back deck from the tail buoy while the streamer is being deployed. (b) One array, consisting of four pairs of airguns, being deployed off the back deck. The cables in this photograph are the air hoses which feed the guns from onboard air compressors. (Photographs courtesy of Schlumberger and Horizon.)

solved by collecting a great multiplicity of data, which is later added or stacked together to improve the ratio of signal to noise. For example, for every 100 seismic records that are stacked together, the signal-to-noise ratio is increased 10 times. A 3D survey covering 10 square kilometres generates an astounding 20 million seismic records, taking up *c.* 320 giga-bytes of computer storage space! The real power of the method is our ability to manipulate an astonishingly large number of records together in order to extract meaningful signals from the Earth. Considerable comput-ing power is required to record and process hundreds of gigabytes of data. Needless to say, large sums of money are involved: a typical 3D survey covering 100 square kilometres costs about a million US dollars to acquire and process.

The final seismic reflection volume is an acoustic scan of the Earth's crust. Compositional layering of rocks at depth generates the laterally con-tinuous reflections that we see on seismic images (Figure 8.2). The size, or amplitude, of these reflections is related to small changes in acoustic velocity and density. These changes are caused by variations in the physi-cal properties of rocks, especially composition and porosity. For example, a limestone is denser and transmits sound waves faster than a more porous and softer mudstone.

We can examine these cubes of data in several different ways. Vertical slices through the cube are similar to geological cross-sections. Since seismic imaging is analogous to echo-sounding, the vertical axis is actually measured in the number of seconds taken to travel down to a particular horizon and back. This 'two-way travel time' can be converted to depth by using information about the speed of sound in the different rock types imaged. Horizontal slices are analogous to geological maps except that fil-tering properties of the Earth cause horizontal resolution to decrease with depth. At 3 kilometres depth, horizontal resolution is *c.* 300 metres and smaller geological features scatter acoustic energy instead of reflecting it. The large volumes of data in a 3D survey require us to use computer work-stations for analysing and interpreting the data. Visualisation software is now used routinely to allow geologists to view the whole data set as a cube, pan through it, rotate and cut it at any angle, and rapidly form a picture of the geology beneath the seabed.

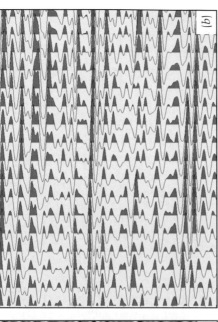

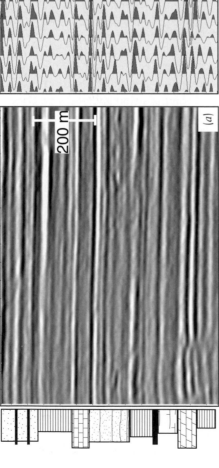

**Figure 8.2.** (a) An example of a seismic profile from a sedimentary basin. Sound waves are reflected from boundaries between layers with different compositions. The continuous reflections that make up the image identify the boundaries where the rock type changes in the crust. In this example, the rocks are flat-lying, rather like a stack of pancakes. To make seismic sections easier to interpret geologists colour-code the amplitudes of the reflections, so that the continuity of a layer boundary is emphasised. (b) The same seismic section with the reflections displayed as waves – it is more difficult to 'see' an image of the rock layers beneath the sea in this version of the section! The column on the left of the figure shows how a geologist might interpret the rock layers from the seismic section.

## 8.4 How do basins form?

About 70 per cent of the surface of the continents is covered in more than 2 kilometres of sedimentary rock. The largest accumulations are in sedimentary basins located on the continental margins and in continental interiors. The size of sedimentary basins varies considerably but a typical example covers several hundred thousand square kilometres. Sedimentary basins are of enormous economic importance: many contain significant quantities of oil, gas and coal as well as minerals. They have acted as major sediment sinks over tens to hundreds of millions of years and contain a unique record of the changes in surface processes (e.g. vertical motions of the Earth's surface, drainage systems, climate) throughout the geological past. Thanks to the efforts of the petroleum industry, most sedimentary basins have been drilled down to depths of at least 4 kilometres. The composition and age of different sedimentary layers are determined from rock chippings collected during drilling. This information provides the geologist with essential data with which to interpret and calibrate seismic images.

The North Sea basin, located between the British Isles and Scandinavia, is a typical example of a sedimentary basin. It contains up to 7 kilometres of infill which mostly consists of layered sand and mud. Before the 1960s no one knew that such a large amount of sediment lay beneath the shallow waters of the North Sea. The origin of this, and other basins, was a puzzle until the advent of high quality 2D seismic data and intensive drilling from the 1960s onwards.

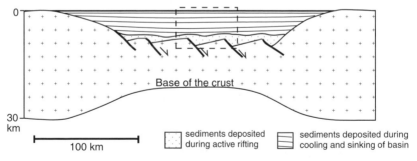

**Figure 8.3.** Extensional sedimentary basins form by stretching or thinning of lithospheric plates. The upper crust stretches by brittle faulting while the hotter, lower crust and lithospheric mantle deform by flowing slowly. Once rifting ceases, the region gradually subsides and fills with sediment. The dashed rectangle indicates the location of Figure 8.4*a*.

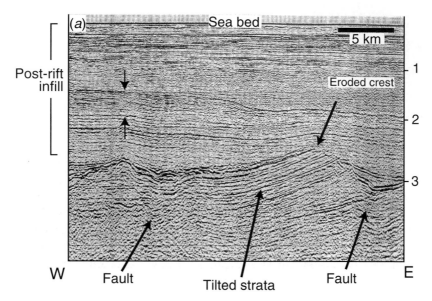

**Figure 8.4.** (a) Seismic profile from the North Sea Basin. The North Sea formed by lithospheric stretching 150 million years ago. Toward the base of the profile, tilted blocks are separated by normal faults spaced 25 kilometres apart. These rotated blocks demonstrate that the Earth's crust has been stretched. The crest of one block shows clear evidence of erosion. Most of the layered sediment that infills the basin was deposited during the cooling phase when stretching had ceased (labelled 'post-rift'). A submarine fan (gently dipping reflections indicated by vertical arrows) developed 50–60 million years ago when sediments were being eroded from the newly uplifted Scottish landmass.

Thirty years on, we now have a good quantitative understanding of sedimentary basin formation. Many, but not all, basins form by stretching or rifting of the Earth's lithosphere, the cold 120 kilometre thick layer which forms the rigid plates. During rifting, the top 10 km of the crust extends by brittle failure on faults (Figure 8.3). Fault-bounded blocks often rotate and look like tilted books on a shelf. The lower crust and the rest of the lithosphere are hotter, and probably deform by flowing slowly. Some sedimentary deposition takes place during rifting and volcanic activity is also common. Once stretching ceases, large-scale cooling and contraction occur and the Earth's surface gradually subsides over a period of around 100

N                                                                                    S

**Figure 8.4.** (*cont.*) (*b*) Silhouette of Piperi, a wedge-shaped island, located between Kithnos and Serifos in the Cycladic Islands, Aegean Sea. It is 200 metres high and represents the uplifted crest of a fault-bounded block exposed about sea level. The steep cliff on the south side of the island is the fault that bounds one side of the tilted block. Note the similarities with the fault block in the seismic profile from the North Sea. Some 150 million years ago fault blocks from the North Sea would have formed similarly emergent islands.

million years. A sedimentary basin is born as this depression grows and fills with sediment transported by rivers from eroding landmasses elsewhere (Figure 8.3).

This simple model is based on observations from areas in the world where this process has happened in the past (e.g. the North Sea) and regions of the Earth's crust that are actively stretching at the present time, such as the Aegean Sea, Greece. Figure 8.4*a* is a typical 2D seismic image from the North Sea basin, which formed by stretching some 150 million years ago. At the bottom of the basin, faulting and tilting of strata are clearly visible. The crests of these tilted blocks were modified by erosion during stretching, when they stuck out above the surrounding seas and formed islands. In the Aegean Sea many wedge-shaped islands protrude above sea level and these islands are the emergent crests of tilted blocks bounded on the steeper side by active faults (Figure 8.4b). In time, erosion will blunt these sharply defined crests. The most striking feature of the North Sea basin is the large thickness of sediment deposited during the later cooling phase. About 4 kilometres of layered muds, sands and chalk blanket the 150 million year old fault blocks (Figure 8.4a).

Seismic surveys in the North Sea are typically recorded to 6 seconds two-way travel time which is equivalent to about 10 kilometres depth and is ideal for studying the geometry of sedimentary layers filling the basin. Adjacent 3D surveys can be merged to give a combined 3D coverage of several thousand square kilometres. The North Sea and other major hydro-carbon provinces such as the Gulf of Mexico now have almost 'wall to wall carpeting' of 3D seismic coverage. Deeper 2D surveys with recording times of up to 15 seconds (c. 40 kilometres) have been acquired to trace faults to greater depths and to demonstrate that the crust beneath the North Sea is thinned as predicted by theoretical models.

## 8.5  How do basins fill?

Basin infill is a large and complex subject and our treatment of it in this chapter is necessarily incomplete. Besides carbonate deposits such as lime-stone and chalk, which are mainly formed from the skeletons and shells of dead sea-life, the most important sources of infill are rivers which carry great quantities of sand and mud from continental interiors into basins and onto continental margins. The modern Orange River in southern Africa pours approximately 200 megatonnes per annum of sediment offshore onto the continental margin, which formed by rifting some 130 million years ago as Africa separated from South America. Seismic data from the Orange Delta reveal the history of sedimentation (Figure 8.5). This delta has grown over the last 60 million years, forming a set of seaward-dipping deposi-tional sequences. These are identified on Figure 8.5b, which is an interpre-tation of the seismic section in Figure 8.5a. Each sequence forms gradually with time as the river load is deposited on the subsiding shelf. In this way, large deltas grow out to sea. In front of the delta, deep-sea channels and sub-marine fans of sediment form. Individual sequences are generated by changes either in the locus of the river mouth, in the rate of sediment supply, or in the relative elevation of the land surface with respect to sea

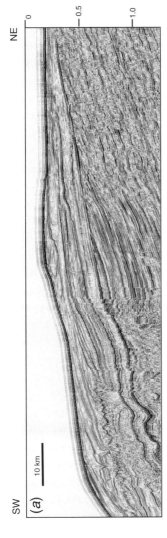
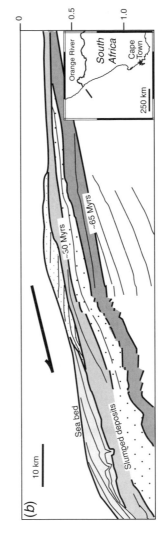

**Figure 8.5.** (a) A vertical seismic profile through sediments deposited at the mouth of the Orange River, southern Africa. (b) An interpretation of the seismic image in (a) showing the individual packages of sediment which built out onto the continental shelf over the last 60 million years. The packages of sediment identified in (b) formed due to changes either in the locus of the river mouth, in the rate of sediment supply, or in the relative elevation of the land surface with respect to sea level through time. (Data courtesy of Namcor and Schlumberger.)

level through time. Deposition is not steady-state: deltas as well as deep-sea submarine fans exhibit pulsing, and the challenge for Earth scientists is to determine how such pulses might be linked to vertical motions of the land surface, changes in sea level, climatic variations or other processes.

Ancient deltas and submarine fans from many locations are seen on seismic images and a good example is illustrated in Figure 8.4a. This North Sea fan is approximately 55 million years old and appears as a series of gently dipping reflections which build eastwards. Such depositional geometries demonstrate that rivers were transporting large volumes of sediment into the slowly subsiding North Sea basin from an uplifted region to the west, which included Scotland.

Figure 8.6 shows another, ancient submarine channel system from the North Sea. A large meandering submarine channel is clearly visible where the cube of data has been cut away (Figure 8.6a). It is imaged because sandy deposits within the channel have different acoustic velocities and densities compared with surrounding deep marine muds. The three horizontal slices are snapshots which show what the channel system looked like at different points in geological time. We can see how a broad and linear set of channels evolved into a single well-established meandering channel within several hundred thousand years (compare Figures 8.6b and c). This channel system was also transporting sediment from an emergent landmass to the west. Combined 3D datasets of this quality can be used to estimate changes in the flux and volume of sediment coming off the adjacent continent through geological time.

## 8.6 Does the solid Earth pulse?

Between 62 and 54 million years ago, an area which includes the British Isles underwent rapid uplift and erosion. Substantial quantities of muds and sands were deposited in the North Sea basin and surrounding regions. Recent seismic imaging shows that submarine fan activity in the North Sea basin waxed and waned every 1–2 million years, in discrete pulses. One of us (White) with Bryan Lovell has suggested that the timing of these sediment pulses is linked to episodic surface uplift related to the injection of magma beneath the British Isles. The timespans of fan deposition and eruption of lavas are very similar and the phase of greatest fan development coincides with the climax of magmatism between 61 and 58 million years ago (Figure 8.7). The remnants of this magmatism can still be seen in western Scotland, and in Northern Ireland, where lavas form the famous Giant's Causeway. Geologists think that these lavas are related to activity of the Iceland plume, a long-lived upward jet of convective mantle beneath the Earth's crust. If correct, White and Lovell's theory is an example of how the sedimentary record can be used to track convection processes and the associated vertical movements of the Earth's surface in our planet's history. The link between mantle convection and sedimentation is unlikely to be straightforward. Nonetheless, a range of independent observations support the hypothesis that the existence of discrete pulses of ancient submarine fan deposition can be linked to convection.

White and Lovell developed these ideas in the North Atlantic region, where large amounts of 2D and 3D seismic data can be used to map submarine fan deposits in considerable detail. The time is now ripe to test this idea in other similar areas where upward and downward movements of mantle occur. One such example is India, which rifted away from the Seychelle Islands 65 million years ago over an upwelling mantle plume. Extensive magmatism occurred, forming the dramatic volcanic landscape of western India (known as the Deccan Traps). Associated uplift and tilting triggered the erosion of large quantities of sediment which were transported by rivers across India into basins off the east coast. Analysis of these sedimentary deposits could help to resolve otherwise inaccessible time-dependent details of the convecting mantle.

## 8.7  A closer look

### 8.7.1  Polygonal faults

Three-dimensional seismic images have been instrumental in discovering a variety of unexpected features which are less easy to recognise in two dimensions. The most obvious examples are submarine channel systems which are barely detectable in cross-section (Figure 8.6). Recently, one of us (Lonergan) with Joe Cartwright described an unusual and extensive network of faults which occur within mud-dominated strata over an area of approximately 150000 square kilometres in the North Sea (Figure 8.8). In cross-section, these faults look like a small-scale version of the faults on Figure 8.4. However, careful 3D mapping has shown that these faults are organised in sets of polygons, rather like gigantic mud cracks. Although they were first discovered in the North Sea Basin, similar polygonal networks have been found in sedimentary basins worldwide. Lonergan and

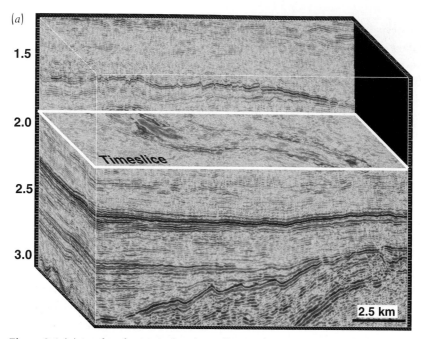

**Figure 8.6.** (a) A cube of seismic data from the North Sea is cut away to reveal the internal geometry of part of the basin. The two vertical sections show a tilted fault block (like those shown in Figure 8.4). The horizontal time slice shows shows a meandering submarine channel which is infilled with sand and surrounded by muddier sediment, 2 kilometres below the present day sea floor. This channel formed approximately 55 million years ago. Sediment within the channel was transported from the west where it was shed from an emergent landmass. Although we can clearly observe the channel on the horizontal slice it is very difficult to identify on the vertical sections. (b) A horizontal slice c. 40 metres higher than previous slice whose position is marked by dashed rectangle. The well-developed meandering channel system, its scale, and the detailed internal geometry can be clearly seen. (c) A horizontal slice c. 250 metres deeper than first slice. At this earlier time, the channel system was broader and more linear. (Data courtesy of Conoco.)

Cartwright do not think that the formation of these fault networks is the result of tectonics on the scale of the lithospheric plates. Instead they suggest that the faults are caused by the way in which water-bearing muds expel fluid during the process of compaction.

(b)

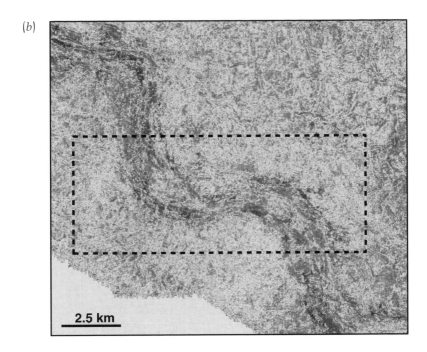

2.5 km

(c)

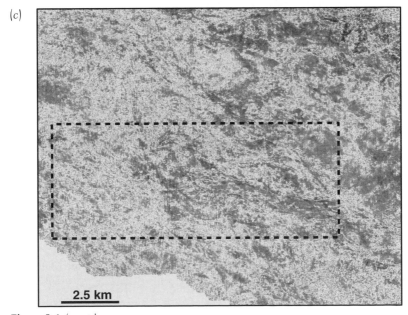

2.5 km

**Figure 8.6.** (*cont.*)

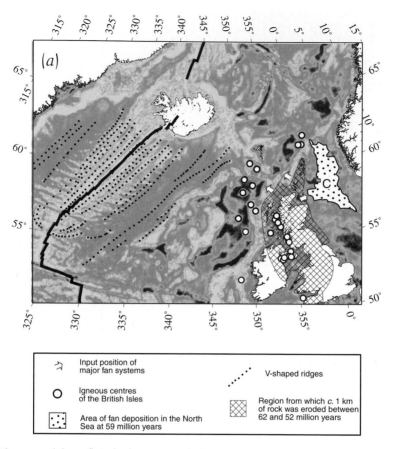

**Figure 8.7.** (a) Sea floor bathymetry map for a region in the North Atlantic Ocean (grey shades indicate shallow water depths and black, deep water). The solid black line shows the position of the plate-spreading axis between the European and North American plates. The present-day mantle plume centre is in southern Iceland. Prominent V-shaped ridges on the sea floor south of Iceland (solid circles) are generated by pulses of hotter mantle moving upwards beneath Iceland and travelling radially away from the centre of the plume. These pulses occur on a timescale of several million years and probably reflect changes within the core of the plume. The white circles around the British Isles are major igneous centres, and the cross-hatched area is a region that underwent uplift and erosion between 62 and 52 million years ago. One of the large submarine fans that formed at the same time in the North Sea is shown in stipple and labelled C.

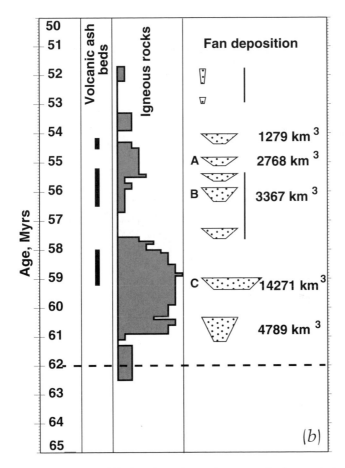

**Figure 8.7.** (*cont.*) (*b*) This chart shows the ages and relative timing of intrusive igneous activity, offshore volcanic ash beds, and submarine fan deposition in the North Sea. The histogram of igneous activity gives an impression of duration and the intensity of magmatism. The stippled wedges represent several different submarine fan deposits (Fan A is illustrated in the cube of seismic data in Figure 8.6, Fan B can be seen on Figure 8.4 and Fan C is in (*a*)). The input of sand into fans was not continuous through time, but occurred in a series of pulses. The horizontal dashed line at 62 million years marks the start of Iceland plume activity.

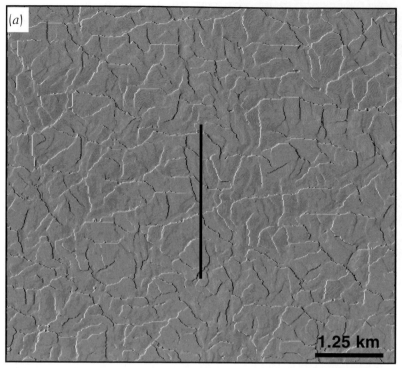

**Figure 8.8.** (*a*) A map of the top of one rock layer in a cube of seismic data from the North Sea, made by tracing a single stratigraphic surface throughout the 3D seismic volume (approximately 1000 lines at 12.5 m spacing). The dip of the mapped surface was then calculated and illuminated by a light source shining from the north west so that the polygonally faulted surface is thrown into relief. (*b*) A vertical seismic profile that illustrates the geometry of the polygonal faults in cross-section. (*c*) A detailed interpretation of (*b*), where the faults are the dashed lines; different rock layers are shown by thin black lines and the surface mapped in (*a*) is shown by a thick black line. (Data courtesy of Fina and Chevron.)

As muds are buried in sedimentary basins, the pore spaces between the clay particles are flattened and even at relatively shallow burial depths these pore spaces no longer link up, thus preventing fluids from flowing between the clay grains. This reduction in permeability means that geologists often consider mudrocks a suitable rock type in which to bury and seal toxic waste. However, the existence of polygonal fault systems in thick mudrock sequences raises questions about the sealing ability of

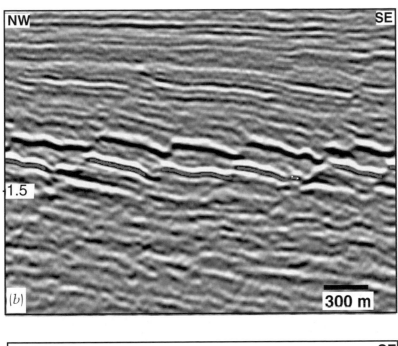

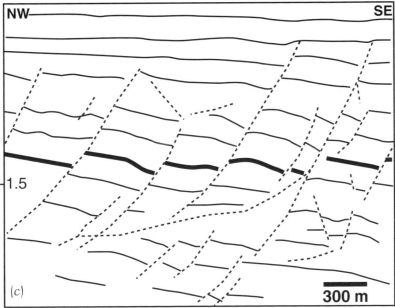

**Figure 8.8.** (*cont.*)

mudrocks. Although fluids will not percolate through the pores between clay grains, they may travel through the fault network.

## 8.7.2  Fluids

As we have implied, seismic data can be used to track fluids in rocks of the Earth's crust. When seismic energy is reflected from the boundary between two rock types, the amplitude of the reflected wave depends upon the acoustic impedances of the two rocks. Acoustic impendance is simply the speed of sound through a given rock multiplied by its density. Impedance depends on the composition of a rock, on its porosity, and on the density and phase of the fluid contained within pore spaces. Ten to twenty per cent of the volume of a typical sandstone consists of pore space between the sand grains and sharp changes in impedance can occur depending on whether the pore spaces are filled with water, oil, or gas. Accordingly, the measured amplitude of a reflection can be used, under favourable circumstances, to identify the pore fluids.

Figure 8.9 illustrates how the amplitude of reflections can help to identify the flow of oil and gas in the subsurface. When oil and gas are extracted from a reservoir, encroaching waters gradually alter the distribution of the remaining hydrocarbons. Successful economic management of a producing field depends upon the ability to predict the changes in such fluid flow. Repeated 3D seismic surveys (known as '4D' or 'time-lapse' surveys) help to monitor the movement of different pore fluids through the subsurface and to discover where remaining pockets of oil or gas lie.

## 8.8 The future

Although 3D surveying is now used routinely by the hydrocarbon industry, its considerable potential has yet to be fully exploited by other scientists. Thanks to the generosity of the petroleum industry, large amounts of 3D imagery from sedimentary basins and continental margins are likely to keep many Earth scientists busy for a long time. In this chapter, we have only described one example which exploits existing 3D seismic databanks: where vertical motions of the Earth's crust are linked to convection within the mantle through analysis of large-scale sediment pulses. There are many other important, and easily addressable, problems which require 3D seismic surveying.

In the oceans, the most obvious target is the mid-oceanic ridge system,

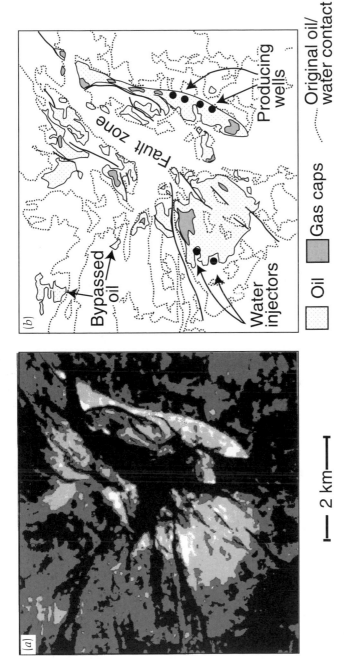

— 2 km —

**Figure 8.9.** (*a*) A seismic amplitude map from a 3D survey of the producing oil field in the Gulf of Mexico. By carefully calibrating the amplitude of recorded seismic waves, Shell geologists discriminated between sediments whose pore spaces contain water (pale grey), oil (dark grey), or gas (white). The extent of the original oil field is gauged from the distribution of oil patches left behind by encroaching waters. Several gas caps are clearly visible, including those which develop at the positions of producing wells. The entire field is dissected by both large and small faults. (Data courtesy of Shell Oil.)

where tectonic plates spread apart and grow, generating large quantities of magma which forms new oceanic crust. Although a considerable number of 2D seismic reflection surveys have shown that molten rock can be imaged at several kilometres depth beneath spreading ridges, we have little understanding of the 3D geometry of these magma bodies. The first 3D survey to target a mid-oceanic ridge was shot during September–October 1997 by Earth scientists from the University of Cambridge and from Scripps Institute of Oceanography, California. They collected data over a 20 kilometre by 23 kilometre area on the East Pacific Rise. These data are now being jointly analysed by the two research groups.

Future targets include active volcanoes located either on land or at sea, active normal faulting in regions that are stretching rapidly, and the accretionary prisms ('mountains' of deformed sediments beneath the sea), which develop where one tectonic plate is subducting beneath another. In each case, there exists the exciting prospect of four-dimensional surveying following eruptions or earthquakes. Such data would undoubtedly help to improve our understanding of the dynamics which govern these fundamental processes.

## 8.9  Further reading

A more technical version of this chapter, with a more complete reference list
   is published in the Millennium Issue of the *Philosophical Transactions
   of the Royal Society*, A. (Lonergan, L. & White, N. J. 1999 Three-
   dimensional seismic imaging of a dynamic Earth. *Millennium Issue. 1.
   Astronomy and Earth Sciences; Phil. Trans. R. Soc. Lond.* A **357**,
   3359–3375.)
Brown, A. R. 1996 *Interpretation of three-dimensional seismic data.* 4th edn.
   Tulsa: AAPG Memoir 42.
Fowler, C. M. R. 1990 *The solid Earth.* Cambridge: Cambridge University
   Press.

# 9
# Geophysical and astrophysical vortices

## N. Robb McDonald

*Department of Mathematics, University College London, Gower Street, London WC1E 6BT, UK (robb@math.ucl.ac.uk)*

## 9.1 Introduction

In 1665, the first year of publication of the world's longest running scientific journal, *Philosophical Transactions of the Royal Society*, Robert Hooke reported his astronomical observations of a 'Prominency of the Belt' in the atmosphere of Jupiter. We now realise, after centuries of speculation and many erroneous theories, that this anomaly is a vast atmospheric storm, or vortex. Known as the Great Red Spot, it continues to swirl coherently and vigorously today, more than three hundred years after publication of Hooke's observations. The ability of the Jovian atmosphere to sustain such a long-lived and well-ordered structure is extraordinary. This is particularly so given our familiarity with the process by which a fluid, be it a gas or liquid, with the freedom to flow in three dimensions may become unstable and flow in an arbitrarily chaotic way. An example of this is the plume of smoke rising from a lit cigarette whose smooth ascent suddenly becomes wildly turbulent. On the other hand, the Jovian atmosphere is constrained to move within a thin[1] shell in which the vertical velocities are negligible compared to horizontal velocities. A remarkable feature of such two-dimensional flows, in stark contrast to the three-dimensional case, is the spontaneous emergence of organised, swirling, smoothly varying flows – vortices – from a sea of turbulence. The Earth's oceans and

---

[1] The actual depth of Jupiter's atmosphere is not well known. However, its dynamics appear to be well modelled by various 'thin-layer' (i.e. shallow water) based theories.

atmosphere are also thin fluid layers and are therefore dynamically similar to Jupiter's atmosphere, and similarly spawn coherent vortices. In our atmosphere, for example, there is the wintertime stratospheric polar vortex, a particularly robust structure in which a region of stratospheric air is effectively isolated from the rest of atmosphere, facilitating the chemistry of ozone depletion. Atmospheric blocking events are large-scale, stationary, very stable vortex structures which disrupt the usual passage of cold fronts and other weather events, leading to sustained periods of unchanging weather for localised regions. Tropical cyclones (also known as hurricanes or typhoons) and tornadoes are dramatic examples of extremely intense atmospheric vortices with sufficiently rapid swirling flows to cause destruction and loss of life for those unfortunate enough to lie in their path. Hurricane Mitch proved to be one of the most devastating weather events of recent times when it smashed into Central America in 1998, claiming over 20000 lives through a combination of extreme winds and heavy rains.

The ocean also has an equally diverse range of vortices. In the Gulf Stream region off the east coast of North America, coherent 'blobs' of both anomalous warm and cold water – Gulf Stream Rings – are observed. These rings, of diameters up to several hundred kilometres, form when the meandering of the Gulf Stream becomes sufficiently contorted to break off and form individual eddies which persist as separate entities for periods of up to six months. Similar vortices are also observed in the Kuroshio Current region off Japan and in the Agulhas Current region at the tip of South Africa. Gulf Stream Rings are examples of vortices which are located near the surface of the ocean, but vortices may exist at any depth. For example, Mediterranean salt lenses (Meddies) are large flat 'discs' of anomalously warm and salty water of Mediterranean origin found in the North Atlantic Ocean. Typically they have diameter of 50km, thickness of several hundred metres, lie about 1000m below the surface of the ocean and have lifetimes of up to two years. In the abyssal ocean, there is good evidence that the dense water formed in the polar oceans is dispersed throughout the global ocean in discrete packets by vortices. On a smaller scale, the coastal surf-zone gives rise to vortices with length-scales from 10 to 100m, which may, for example, cause rip currents.

The ability of ocean vortices to transport heat, salt and momentum over large distances, together with their anomalous biological and chemical properties, make them important components of global ocean circulation. For example, Agulhas eddies are responsible for the transport of at least $2.2 \times 10^{20}$ J yr$^{-1}$ of heat and at least $1.4 \times 10^{13}$ kg yr$^{-1}$ of salt from the Indian to the Atlantic Ocean, and thus form an important link in the global 'conveyor belt' that is the world's ocean circulation. Meddies serve to spread significant amounts of relatively salty water from the Mediterranean throughout the North Atlantic. How such vortices redistribute heat and salt throughout the global ocean determines, to a large extent, our climate. Unfortunately, since typical diameters of ocean vortices range from tens to hundreds of kilometres, they are difficult, if not impossible, to resolve with present day climate models. Thus, understanding their behaviour is a key aspect in predicting the effects of global warming.

The Great Red Spot is the most well-known astrophysical vortex. Its dimensions are huge: 24 000 km long and 11 000 km wide (compare with the Earth's average radius of 6371 km), and it has wind speeds of 100 m s$^{-1}$ swirling about its centre. In common with Meddies, it is also thought to be relatively thin, so that it is also 'disc-like'. In fact, this 'disc' or 'pancake' description is a good image to bear in mind for the shape of most vortices described in this article. In addition to the Great Red Spot, there are many other long-lived vortices in the atmospheres of Jupiter and the other giant planets, namely Saturn and Neptune. Most notably, Neptune has the Great Dark Spot, another huge atmospheric vortex located, curiously, at almost the equivalent latitude as the Great Red Spot on Jupiter. It is extraordinary that 90 per cent of these astrophysical vortices are anticyclonic (including both the Great Red Spot and the Great Dark Spot), i.e. their winds circulate in the opposite sense to the rotation of their host planet. On the other hand, here on Earth most of the atmospheric vortices that give us our day-to-day weather tend to be cyclonic. Yet in the ocean, as in astrophysical vortices, there is a bias toward anticyclones over cyclones. In fact, our ocean is, in many respects, more akin to the atmospheres of the giant planets in that its vortices are small compared to the extent of the surrounding fluid, i.e. they exist in a spacious environment. Earth's atmospheric vortices, on the other hand, are of planetary scale and are therefore of a much larger size relative to their surrounding fluid.

It is not surprising that geophysical and astrophysical vortices have received considerable attention from oceanographers, meteorologists, planetary and climate scientists. They are also of considerable appeal to theoreticians working in fluid dynamics. The nonlinear equations of motion of fluid dynamics (here, the 'shallow water equations') are capable of representing highly chaotic and turbulent flow. It is surprising that the same equations give rise to organised, almost circular, vortex behaviour with smoothly varying flow properties. Moreover, these vortices are extremely long-lived, resisting break-up into smaller-scale turbulence. As already noted, the emergence of large-scale vortices from a sea of turbulence is unique to thin-layer (shallow) fluids in which the flow is (or almost is) two-dimensional. This is quite different from fluid with the ability to flow freely in three dimensions, where turbulence continually breaks up into ever smaller scales. The ability of modern computers to perform rapid computations has proved a particularly useful tool in studying the complex process by which two-dimensional fluids organise themselves into larger and larger scales via vortex merger. In view of this complexity, it is fascinating to note that the vortex merger process was described as early as the nineteenth century, with great insight and accuracy, by Edgar Alan Poe in his fictional story *A Descent into the Maelström* in 1845.

A fundamental and important aspect to the study of geophysical vortices is the prediction of their trajectories. For instance, it is widely believed that global warming will lead to increase in the frequency of topical cyclones. Accurate prediction of their likely paths as they evolve could possibly save many lives. In the ocean it is vital to predict where, in what quantities, and how rapidly vortices carry heat and salt. Given the many different factors that may influence vortex motion, the trajectory prediction is a very challenging problem. For example, it seems certain that thermodynamics plays a significant role in the motion of tropical cyclones. Thermodynamics also plays a major role in the dynamics of Gulf Stream Rings, along with other effects such as the presence of neighbouring rings, the Gulf Stream itself, bottom topography and friction. An effect important to all but the smallest of vortices (such as tornadoes – see later discussion in Section 9.2) is that of the rotation of the Earth. In addition to the restriction of vertical velocities imposed by the thinness of geophysical and astrophysical fluids, rotation imparts a rigidity, or elasticity, to the fluid in the north–south direction. As will be discussed in Sections 9.2 and 9.3, this elasticity allows the existence of large-scale oscillations of the fluid in the north–south direction, resulting in waves travelling in the east–west direction. These are known as Rossby waves and are of fundamental importance to the dynamics of geophysical and astrophysical fluids. Perhaps most remarkable is that a planet's rotation provides a mechanism for vortices to self-propel or drift: they are not simply pushed about by the prevailing winds or currents, but are able to move by themselves. It is this self-propulsion phenomenon that this article will concentrate on. For example, even in the absence of external currents, Meddies travel at a speed of the order of several centimetres a second. Over a typical lifetime of two years, this amounts to displacements of the order of 2000 kilometres.

Our understanding of the self-drift mechanism of vortices has come through a combination of mathematical and numerical models together with laboratory experiments and observations of real-world vortices. This chapter aims to explain the fundamental physics involved in the motion of vortices. Recent results and outstanding problems will also be discussed. Throughout, the focus will be upon monopole vortices, i.e. vortices comprising fluid which circulates in one direction only. Observations show that monopoles are the most commonly occurring type of vortex, but it is important to note that there is observational evidence for dipoles (i.e. two counter-rotating vortices), of which atmospheric blocking events are one such example. There are many important aspects in the study of geophysical vortices that will not be addressed here; for example their genesis, stability, and the effects of thermodynamics and friction.

## 9.2  Fundamentals: some effects of rotation

We are familiar with the Earth's rotation, taking 24 hours to make one complete revolution. In fact, all the planets in the Solar System rotate, Jupiter taking about 10 hours to make one revolution. While planetary rotation is not directly responsible for the formation of vortices (vortices would be perfectly capable of forming on a non-rotating planet), the vast scales and longevity of many vortices mean that planetary rotation must play a significant role in their dynamics. In particular, rotation gives rise to Coriolis forces, and, furthermore, imparts a rigidity, or elasticity, to the fluid in the poleward direction. These fundamental effects are discussed in this section.

### The Coriolis force

Despite their very different scale and composition, the atmosphere and oceans of Earth are dynamically similar to Jupiter and, for that matter, all the other planets. They are all thin layers of fluid (gas or liquid) and are governed by the same laws of motion, namely Newton's laws. When describing the flow of such fluids, careful consideration must be given to the state of motion of the observer. It is natural for us to describe the dynamics of the atmosphere and oceans from the point of an observer fixed relative to the surface of the Earth. Such an observer will obviously describe the dynamics of, say, a moving particle differently from an observer who is not rotating, since the particle will behave differently according to each observer. To account for this difference, it is necessary for the observer fixed to the planet to include an extra force, namely the Coriolis force. A particle moving tangentially to a planet's surface (i.e. horizontally) experiences a Coriolis force, $F_{\text{Coriolis}}$, which also acts horizontally.[2] It has magnitude equal to twice the product of the momentum of the particle $mv$, where $m$ is the mass of the particle and $v$ its velocity, with the component of the planet's angular velocity vector in the direction of the local vertical. This component varies with the sine of the latitude $\phi$ and so is a maximum at the poles and vanishes at the equator, see Figure 9.1a. Thus, $F_{\text{Coriolis}} = 2\Omega mv \sin\phi$ where $\Omega$ is the angular velocity of the Earth. The quantity $2\Omega \sin\phi$ depends only on the location and speed of planetary rotation and is frequently referred to as the Coriolis parameter $f$. Importantly, the direction of the Coriolis force acting on the particle is also in the horizontal plane, but at right angles to the direction of the velocity, see Figure 9.1b. In particular, in the northern (southern) hemisphere it acts to the right (left) looking in the direction of motion.

## The Rossby number

The effect of the rotation of a planet on a vortex can be gauged through the magnitude of the Rossby number $Ro$. This measures the relative importance of the strength of the swirling flow of the vortex to the Coriolis force and is given by $Ro = V/fL$. Here $V$ is a typical swirl velocity associated with the vortex, $f$ is the Coriolis parameter and $L$ is the horizontal extent of the vortex. Roughly speaking, if $Ro$ is of the order of unity or less, then rotational effects are significant in the vortex dynamics. For example, typical values for Agulhas rings at latitude $35°\text{S}$ are $f = 8 \times 10^{-5}\text{s}^{-1}$, $V = 50\,\text{cm}\,\text{s}^{-1}$ and $L = 80\,\text{km}$, giving $Ro \cong 0.1$, indicating that the Earth's rotation plays a significant role in its dynamics. On the other hand, tornadoes with their large velocities and relatively small size have large $Ro$, so that the Earth's

---

[2]  There is also a vertical component to the Coriolis force, but for geophysical and astrophysical fluids this is usually small in comparison to buoyancy forces and so will be ignored here.

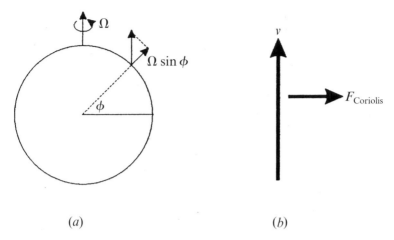

$(a)$                                        $(b)$

**Figure 9.1.** $(a)$ The variation with latitude of the local vertical component of the Earth's rotation vector $\Omega$. The projection of the vector onto the local vertical varies as the sine of the latitude $\phi$, and is proportional to the local strength of the Coriolis force. The Coriolis force therefore decreases from a maximum at the poles to zero at the equator. $(b)$ The direction of the Coriolis force $F_{\text{Coriolis}}$ is such that, in the northern hemisphere, it acts to the right of the direction in which the particle with speed $v$ is moving. In the southern hemisphere the Coriolis force reverses direction and acts to the left.

rotation plays only a minor role in their dynamics. Likewise, the familiar 'bath-tub' vortex has a Rossby number of at least $10^3$, and therefore the Earth's rotation is insignificant in its dynamics. This destroys the myth that, in general, water will disappear down a sink in opposing senses according to which hemisphere it is in. In fact, the sense of swirl of the draining water is much more sensitive to factors like the motion of the fluid just prior to pulling the plug or the shape of the bath. Indeed, it is a simple matter to make the water disappear down the sink in any direction you wish simply by giving the water a swirl with your hand in the desired sense immediately before pulling the plug.

To a large extent, the motions in both the Earth's atmosphere and oceans are small Rossby number flows and are in a state of *geostrophic* balance. That is, the Coriolis force balances the horizontal pressure gradient. Pressure gradients arise due to the presence of both high- and low-pressure cells (referred to as anticyclones and cyclones respectively). Given that fluid tends to move from regions of high pressure to low pressure, as it does so it will experience a Coriolis force. As a consequence, in the northern hemisphere, fluid tends to spiral clockwise as it moves out from a high pressure cell and spiral anticlockwise as it moves towards a region of low pressure. Therefore, in the northern hemisphere cyclones have anticlockwise circulation and anticyclones have clockwise circulation. The senses of the circulations are reversed in the southern hemisphere because the Coriolis force changes direction.

## The conservation of potential vorticity

At the heart of understanding the dynamics of geophysical and astrophysical fluids is the conservation of potential vorticity $Q$. This fundamental law is derived from the equations of motion for a shallow layer of fluid on a rotating planet in the absence of friction. Its precise form depends on the details of the fluid model being considered. The main points can be illustrated by considering a single layer of homogeneous fluid, in which case the potential vorticity is $Q = (f + \omega)/h$. Here $h$ is the depth of the fluid layer which may vary owing to, for example, topography, and $\omega$ is the relative vorticity. The relative vorticity measures the local circulation or 'spin' of the fluid about the vertical: positive relative vorticity implies local anticlockwise (clockwise) circulation in the northern (southern) hemisphere and negative relative vorticity implies local clockwise (anticlockwise) circulation in the northern (southern) hemisphere. So powerful a concept is the potential vorticity $Q$, that knowledge of its distribution everywhere enables the velocity field to be calculated, which in turn can be used to move and redistribute the potential vorticity, and so on. Thus the potential vorticity field controls the dynamical evolution of the fluid and it is used extensively in modern weather forecasting models.

As mentioned previously, the presence of rotation imparts a 'stiffness' to the fluid in the meridional direction (i.e. the direction of changing latitude $\phi$ or north–south direction). More generally, this 'stiffness' is in the direction of the gradient of $Q$. As an example, consider the simple case where variations in $Q$ due to latitude may be ignored in comparison to variations in topography. In this case, for a fluid which initially is still (i.e. $\omega = 0$) $Q \cong f_0/h$, where $f_0$ is the now constant Coriolis parameter. In particular, consider a fluid of constant shallow depth for $y > 0$ and constant, but deeper, depth for $y < 0$, see Figure 9.2. Thus the line $y = 0$ separates two regions of differing $Q$, with higher constant $Q$ for $y > 0$ and lower constant $Q$ for $y < 0$. If the interface is perturbed as shown in Figure 9.2, the conser-

shallow

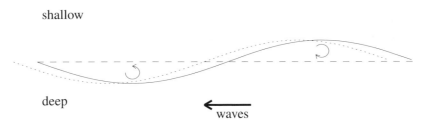

deep

← waves

**Figure 9.2.** The Rossby wave mechanism. The dashed line along $y=0$ separates shallow $(y>0)$ from deep $(y<0)$ water. The solid line represents an initial perturbation to the line of fluid particles which initially lay over the escarpment running along $y=0$. Fluid moving from shallow to deep water acquires positive relative vorticity resulting in a local anticlockwise circulation as shown by the curved arrow. On the other hand, fluid moving from deep to shallow water produces negative relative vorticity resulting in a localised clockwise circulation. The effect of these circulations is to shift the initial perturbation to a new position as shown by the dashed line. As the process evolves the perturbation travels to the left as a wave.

vation of potential vorticity implies that fluid that has moved from deep to shallow fluid will acquire negative relative vorticity $\omega$ to compensate for the decrease in depth $h$ and therefore starts to circulate clockwise. On the other hand, fluid that has moved from shallow to deeper fluid acquires positive relative vorticity $\omega$ and circulates anticlockwise. The principle giving rise to these circulations as fluid columns are stretched when they change depth is the same as that by which a spinning ice-skater is able to spin up or down by drawing in or extending their arms. The effect of the circulations is to cause the disturbance to move to the left. The conservation of potential vorticity implies the existence of such waves whenever a gradient exists in the background $Q$ field. Variable topography is only one factor which may lead to variation in the potential vorticity. Equally, variation in $Q$ can arise owing to the ubiquitous meridional variation in $f$ (i.e. the sphericity of the planet), or due to the presence of currents having spatially varying strength, i.e. variations in $\omega$ such as may occur near the Gulf Stream or the strong latitudinal jets of Jupiter. In any case, no matter what causes the variation in the background potential vorticity, the consequence is the fluid medium is able to support waves. Such waves are called Rossby waves and are always such that in the northern hemisphere they travel with higher values of $Q$ on their right facing in the direction they are travelling.

The existence of Rossby waves has important implications for the motion of vortices, since any translating vortex must disturb the background $Q$ field which, in turn, leads to the radiation of Rossby waves. Analogous to a ship moving on the water's surface, such radiation will lead to a loss of energy, or drag, on the vortex. The consequences of this radiation, and models to study its effect, will be discussed in Sections 9.3 and 9.4.

## The westward drift of a vortex

In the absence of variable topography and ambient winds, currents and jets, the variation in the Coriolis force with latitude causes vortices to self-propel in the westward direction; or, more precisely, in the opposite sense to the planet's rotation. In effect the curvature of the planet provides the engine by which vortices drift. Figure 9.3 shows a vortex consisting of an isolated lens of fluid spinning about its centre. In the northern hemisphere, associated with such a vortex is a clockwise circulation, i.e. it is an anticyclone. As fluid swirls within the vortex it experiences a Coriolis force: as fluid moves eastward in the northern half of the vortex there is a net Coriolis force directed southward, and as fluid moves westward in the southern half of the vortex the Coriolis force is northward (see Figure 9.3a).

**Figure 9.3.** (opposite) The westward drift mechanism for an isolated lens of fluid (an anticyclonic votex) (a) and (b); and a cyclonic vortex (c). In (a) there is a Coriolis force on the fluid as it moves within the vortex. The Coriolis force on the fluid when moving from left to right in the northerly part of the vortex is directed to the south, and when moving from right to left in the southerly part of the vortex is directed to the north. The northerly component of the Coriolis force is less than the southerly component owing to the latitudinal variation in the Coriolis force (see Figure 9.1). It follows that there is a net southward force on the vortex. Since the motion is steady, this force is balanced by an equal and opposite force. In (b) the vortex drifts as a whole to the left resulting in a northward Coriolis vortex acting on the whole vortex which precisely balances the net southward force in (a). The flow exterior to the vortex must be considered for the case of a cyclone. In (c) it is shown that as the cyclone drifts in the east–west direction it pushes surrounding fluid either to the north or south. Fluid that has been pushed south produces positive relative vorticity and hence anticlockwise circulation. Fluid pushed to the north produces negative relative vorticity and hence clockwise circulation. The combined effect of the surrounding circulations is to push the vortex to the west.

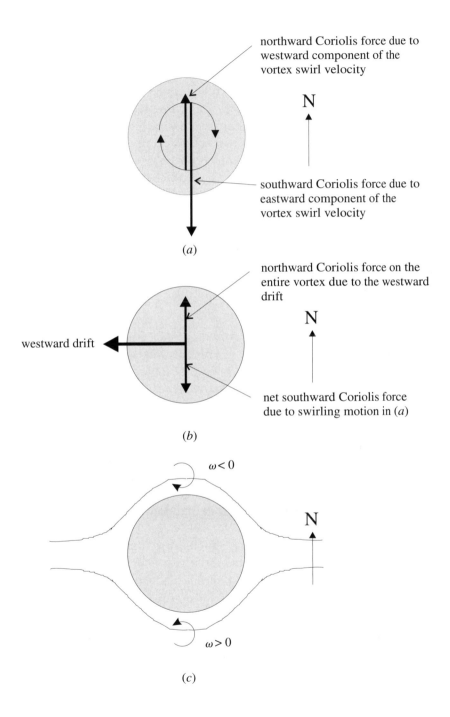

northward Coriolis force due to
westward component of the
vortex swirl velocity

N

southward Coriolis force due to
eastward component of the
vortex swirl velocity

(a)

northward Coriolis force on the
entire vortex due to the westward
drift

N

westward drift

net southward Coriolis force
due to swirling motion in (a)

(b)

$\omega < 0$

N

$\omega > 0$

(c)

Since the magnitude of the Coriolis force increases in a northward direction, the Coriolis force experienced by the top half of the vortex is greater than that experienced by the bottom half of the vortex. This is indicated in Figure 9.3a by the differing length of the north–south arrows. Thus, this motion alone leads to a net southward force acting on the vortex. However, if the vortex motion is steady, there can be no net force on the vortex (otherwise, by Newton's laws, the vortex would accelerate). Therefore the vortex must drift westward as a whole (Figure 9.3b) in order to provide a countering northward Coriolis force required to balance the force due to the swirling motion in Figure 9.3a. Typically the horizontal extent of the vortex is small and so the westward drift of the vortex is correspondingly small compared to the swirling velocity within the vortex itself.

A similar argument to that above applied to cyclones in the northern hemisphere which have anticlockwise circulation leads to the erroneous conclusion that they drift eastward. However, the situation depicted in Figure 9.3a,b is applicable only to vortices which consist of isolated lenses of fluid and, as such, are always necessarily anticyclones. Anticyclones correspond to anomalous high pressure and, for a single layer of fluid, this consists of a local thickening of the fluid, which decreases with distance from the central pressure, or fluid thickness, maximum. If this thickness decreases to zero the vortex is an isolated 'blob' of fluid with finite volume. On the other hand, cyclones consist of an anomalous region of low pressure or, equivalently, a local depression in a layer of fluid. Since the layer thickness increases with distance from the central pressure minimum, a cyclone cannot be isolated in the sense that it cannot be of finite volume. Therefore fluid exterior to the vortex can no longer be ignored and its dynamics must be considered. Figure 9.3c shows the situation for the case of a cyclone in the northern hemisphere. While the argument outlined above still applies to the core of the vortex (shaded), fluid external to this core must, as demanded by the conservation of potential vorticity, acquire relative vorticity as it pushes around the vortex. Fluid forced northward increases $f$ and so acquires negative $\omega$, hence producing a local secondary clockwise circulation. Similarly, fluid forced south acquires positive $\omega$, hence producing a local secondary anticlockwise circulation. The net effect of these secondary circulation cells surrounding the cyclone is to advect the cyclone west. Thus there are two competing effects: that shown in Figure 9.3b which, in the case of a cyclone tends to cause eastward drift, and the effect in Figure 9.3c producing westward drift. Careful mathematical analysis of the governing shallow water equations shows that this latter effect always wins, and cyclones, like anticyclones, always drift west, albeit at a speed less than that of anticyclones. Note that westward drift for both cyclones and anticyclones also occurs in the southern hemisphere.

It is an illuminating exercise to account for the westward drift of a vortex on a rotating planet from the point of view of an inertial observer, i.e. one that is not fixed to the rotating planet but is instead stationary. Crucial to the previous argument for westward drift was the existence of Coriolis forces. This immediately poses a problem since an inertial observer cannot appeal to Coriolis forces, since such forces exist only for an observer who is rotating. What, then, causes this westward drift? By considering the vortex to be a solid spinning object, it has recently been demonstrated that the vortex drift obeys the same equations as a child's spinning top. The westward drift turns out to be analogous to the precession of the spinning top. Precession is the slow circular motion about a fixed vertical axis of a rapidly spinning body. According to this observer's mathematical analysis, the vortex precesses about a planet's rotation axis at a speed slightly less than the speed of rotation of the planet. Thus, relative to the planet, the vortex appears to drift to the west.

## 9.3 The early life of a vortex

In the previous section it was argued that a vortex in *steady* motion must necessarily move westward at a constant speed. This poses the question of how, if at all, does a vortex reach a steadily moving state? Much research has been aimed at answering the question of how a vortex evolves from an initially stationary state. This question is relevant, for example, to tropical cyclones where a localised intense low-pressure cell forms, quickly establishing a rapid swirling flow about its centre. How, subsequently, does the storm evolve? Where does it go and how quickly does it move? Mathematical modelling, supercomputer simulations and laboratory experiments have all been used to tackle this problem.

A common approach has been the use of so-called *quasigeostrophic* theory. This is an approximation to the shallow-water equations and is valid only for small Rossby number *and* when variations in the thickness of the layer of fluid are small compared to the average thickness of the layer, i.e. the surface of the fluid is almost flat. This latter requirement precludes the study of the isolated 'blobs' of fluid (as used in the discussion in Section 9.2) since these necessarily have large variations in layer thickness. Importantly, the use of quasigeostrophic dynamics still retains the essential nonlinearity required to study vortices but yields a much simpler mathematical system than the more general shallow-water equations. For example, in quasigeostrophic theory it is possible to express the dynamics concisely in one equation for one variable. Contrast this with the significantly more onerous task of solving the shallow-water equations: a complex system of three nonlinear, coupled equations in three unknowns. Though the small Rossby number assumption implicit in quasigeostrophic theory is acceptable for many geophysical vortices, the further assumption requiring small deviations in the layer thickness is, unfortunately, not so readily justifiable, particularly for ocean vortices. Nonetheless, much can be, and indeed has been, learned about the behaviour of vortices within this framework.

Clear from previous studies is that the first stage of the evolution of the vortex is the development of a secondary dipole circulation known as the $\beta$-gyres. A dipole is a pair of counter-rotating vortices which mutually push each other in the direction along the line bisecting them, thus providing a mechanism for self-propulsion. Figure 9.4 illustrates how they form as a direct consequence of potential vorticity conservation. The anticlockwise circulation associated with a northern hemisphere cyclone induces a northward (southward) motion to its east (west). Recalling (cf. Figure 9.3c) that northward displacement of fluid results in negative relative vorticity production and southward displacement results in positive relative vorticity production, it follows that clockwise circulation develops to the east of the vortex and anticlockwise circulation develops to the west, i.e. a dipole is generated. The sense of the circulations in each half of the dipole is such that, initially, it moves northward (Figure 9.4a) moving the cyclonic vortex with it as it goes. In turn, the circulation of the vortex rotates the axis of the dipole anticlockwise as shown in Figure 9.4b. Thus the cyclone moves northwestward along a curved trajectory. A similar argument shows that a northern hemisphere anticyclone moves southwestward. In general the $\beta$-gyres are relatively weak vortices compared to the vortex itself and so the translation velocity of the vortex is small compared to a typical swirl velocity. The role of the $\beta$-gyres has been confirmed in laboratory experiments and by solution of the equations of motion using powerful computer-based methods.

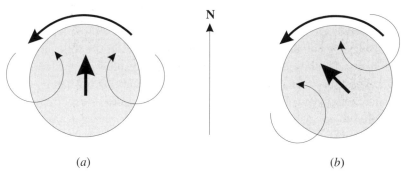

$(a)$ $(b)$

**Figure 9.4.** Formation of the $\beta$-gyres for a cyclonic vortex (shaded). Shown in $(a)$ is the situation immediately after the vortex swirling flow (the anticlockwise flow of the vortex is indicated by the large curved arrow to the north of the vortex) has been switched on. The vortex flow causes fluid east of the vortex centre to move northward, creating negative relative vorticity and hence the local clockwise secondary circulation shown. To the west of the vortex, the vortex flow causes fluid to move south, creating positive relative vorticity and hence the local anticlockwise secondary circulation shown. The effect of the secondary circulations is to move the cyclonic vortex in the direction indicated by the short arrow centred on the vortex, i.e. north. The situation a short time later is shown in $(b)$. Here the main vortex circulation has shifted the smaller secondary circulation around anticlockwise, which now moves the vortex in a northwest direction.

The next stage in the evolution is less well understood. This is because after times of the order of the time taken for the vortex to move a distance comparable to its width, the effects of Rossby wave radiation become important. On the Earth this corresponds to about 2–3 days for an atmospheric vortex and about 10 days for an oceanic vortex. Recall that the meridional gradient in the Coriolis parameter imparts elasticity to the fluid enabling large-scale waves to travel in the east–west direction, known as Rossby waves. As the $\beta$-gyres cause the vortex to translate, the resulting disturbance to the background potential vorticity field leads to the radiation of Rossby waves, in analogy with the way that a moving ship generates surface waves. It is worth emphasising, however, that Rossby waves are quite different to surface water waves. For instance, they are of a much larger scale, travel only in the east–west direction and do not require the presence of a deformable free surface for their existence. The radiation of Rossby waves poses a difficult problem for theoreticians: the

Rossby wave wake must drain energy from the vortex, thus affecting its intensity and structure, and this, in turn, must influence the form of the Rossby wave wake. In most cases the energy carried in the wake is only a small fraction of the energy of the vortex and so the effect is weak and may be considered as a small perturbation to the vortex. Nevertheless, wave radiation may have considerable effect over large times. Most attempts to study this stage of the evolution assume that the vortex is translating almost steadily in a mainly westward direction. Calculating, for example, the energy flux in the wake and relating it to the rate of change of the vortex energy, enables the response of the vortex to be calculated. Results reveal that the effects of radiation include a slow meridional drift in the vortex and a slow shrinkage of the vortex radius. There are, however, some technical problems with the 'almost steady' assumption, and the effects of Rossby wave radiation on vortex motion is an ongoing area of active research.

While thorough understanding of the evolution of the vortex can only come through the formulation and solution of simplified mathematical models, it is clear this is very difficult task. Fortunately the use of high-speed computers using sophisticated algorithms has proved an invaluable tool, helping to shed light on many aspects of vortex behaviour. Recent progress has been made in the high-resolution numerical modelling of quasigeostrophic vortices. Recently David Dritschel and co-workers at the Universities of Cambridge and St Andrews have developed the Contour-Advective Semi-Lagrangian (CASL) algorithm suitable for use on high-speed computers to model the evolution of quasigeostrophic vortices. The CASL algorithm is based on potential vorticity conservation and, unlike previous numerical methods, efficiently handles both the patch of constant vorticity representing the vortex and the continuous variation of the background potential vorticity. Their results demonstrate two principal processes in the vortex evolution: (i) the redistribution of the background potential vorticity by the vortex and (ii) the deformation of the vortex boundary. They also show that the westward drift speed increases with increasing strength of the vortex. Moreover, the meridional motion is shown exclusively to be a result of the trailing wake of the vortex and is a maximum for a vortex neither too intense nor too weak. Figure 9.5 shows results from a typical numerical experiment on an evolving vortex using the CASL algorithm. The potential vorticity field, which increases uniformly in the northward $y$-direction (thus representing the meridional

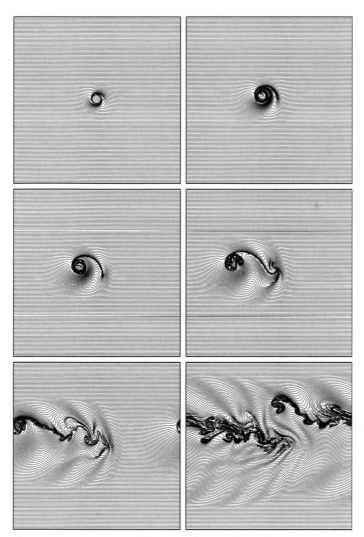

**Figure 9.5.** Computer simulation evolution of a moderate intensity, quasigeostrophic cyclonic vortex on the $\beta$-plane, obtained using the CASL algorithm. Contours of potential vorticity taken at 1, 6, 11, 21, 41 and 81 vortex rotation times are shown. The circulation of the vortex 'winds-up' the contours of potential vorticity and moves northwest due to formation of the $\beta$-gyres (see Figure 9.4). The flow becomes ever more complex as the vortex begins to lose coherence after long times. The Rossby wave wake behind the moving vortex is visible. North is vertically upward and the domain is periodic in the east–west direction.

variation in the background potential vorticity of a spherical planet) and a circular patch of uniform vorticity are shown. As predicted by the $\beta$-gyre argument the vortex, being cyclonic, drifts northwest. The effects of the Rossby wave radiation are clear, as manifested by the undulating potential vorticity contours in the far-field. Of particular importance is the shedding of vorticity in the 'turbulent wake' of the vortex which has a significant effect on, for example, the vortex trajectory and speed. Such an effect, while clearly important, has not been modelled in any mathematical models to date. Its incorporation in models is another challenge facing theoreticians.

Stimulated by theory and real world observations, laboratory experiments have been successfully used to study vortices. Figure 9.6 shows an example of a laboratory experiment. Here a vortex moves in a fluid in which the depth decreases uniformly in the positive $y$-direction. This implies an increase in the background potential vorticity in the same direction and mimics the meridional variation in the ambient potential vorticity on a spherical planet. Thus the direction of increasing $y$ is the laboratory analogue of true north. In addition, the whole aparatus is mounted on a rotating table to simulate planetary rotation. Experimentally, a vortex is created by sucking out water locally through a tube, or vigorous stirring of a localised region. The vortex itself is dyed so that its progress can be observed. Of particular note is the wake developing behind the vortex, similar to that observed in the numerical experiment shown in Figure 9.5. Another feature of the experiment shown in Figure 9.6 is the inclusion of a sudden change of depth midway along the tank, as indicated by the vertical dark line. That is, there is an escarpment running north–south separating shallow water on the right from deep water on the left. Initially the vortex drifts to the 'west' until it reaches the escarpment where, remarkably, it appears to be reflected. This is due to dipole formation as anticyclonic vorticity is produced as fluid is forced by the vortex to move from deep to shallow water. The dipole effect is sufficiently strong to overcome the westward drift due to the $\beta$-gyres. Such vortex–topography interaction is of great interest and is, for example, thought to occur frequently in the ocean and is the subject of much current research interest.

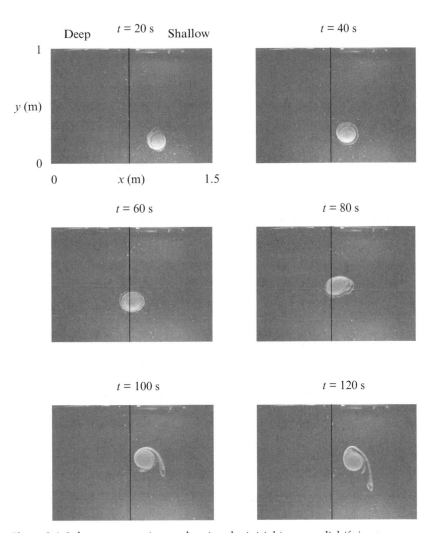

**Figure 9.6.** Laboratory experiment showing the initial 'westward' drift (up to $t = 60$ s) of a cyclonic vortex. The vortex is made visible by marking it with dye. The fluid depth decreases uniformly with $y$ and this corresponds to the 'northward' direction. In this experiment there is an escarpment (i.e. a sudden change in depth) running 'north–south' (indicated by the dark vertical line) separating deep and shallow regions. Extraordinarily, the vortex appears to reflect when it reaches the escarpment at $t = 80$ s and travels back toward the 'east'. This is due to dipole formation as anticyclonic vorticity is produced as fluid is forced by the vortex to move from deep to shallow water.

## 9.4 Future developments in vortex modelling

It is clear that the task of modelling vortices in geophysical and astrophysical environments is a demanding one. Even in the relatively simple problem of quasigeostrophic vortex evolution, there is still much to understand. As the numerical experiments using the CASL algorithm demonstrate, the shedding of vorticity in a 'vortex sheet' plays a significant role in their evolution. The fact that the meridional velocity is maximal for intermediate strength vortices is interesting and indicates that in this case the vortex and the vorticity in the trailing vortex sheet pair up as a dipole to drive significant meridional motion. This is also consistent with recent numerical experiments for vortices concerning vortex interaction with simple forms of topography in the form of an infinitely long escarpment or a seamount. They show that, when the strength of the vortex is comparable to the strength of the background potential vorticity gradient, dipole formation appears to be a ubiquitous, dominant and robust feature in the overall dynamics. Indeed, it is responsible for the 'reflection' phenomenon shown in Figure 9.6. Of particular interest to the theoretical modellers is whether it is possible to analytically model the interaction of this vortex sheet with the vortex? The difficulty here is that, in the interesting case when the vorticity in the wake is comparable to the vorticity of the vortex, their interaction is nonlinear and, moreover, the wake energy remains localised near the vortex and is not radiated away to infinity as it is for Rossby waves. Of course, a successful model is desirable in order to gain quantitative predictions for the vortex trajectory, in particular its meridional displacement.

Perhaps the greatest immediate challenge to the modelling community comes from the need to relax the restrictions imposed by the quasigeostrophic assumption. Ocean eddies in particular have variations in layer thickness sufficiently large as to invalidate quasigeostrophic theory. There is then a need to use more appropriate dynamical theories using the shallow-water equations for example, which are valid for arbitrary variations in the thickness of the fluid layer. From a numerical modelling point of view, there has been recent progress in developing the CASL algorithm for the more general shallow-water equations offering the exciting possibility of the high-resolution study of shallow-water (as opposed to quasigeostrophic) vortex evolution. An important property of the shallow-water equations is that there is an asymmetry – not present in the quasigeostrophic equations – in the behaviour of cyclones and anticyclones. This asymmetry may explain the observed bias of the Earth's ocean and the atmosphere of Jupiter toward anticyclones over cyclones. In fact it is straightforward to argue, using the shallow-water equations, that cyclones will radiate Rossby waves on the $\beta$-plane whereas anticyclones will not. How this decay is manifested, and over what time it takes require further study, using both mathematical and numerical modelling.

Most of this chapter has been concerned with the behaviour of vortices comprising a single layer of fluid. Many vortices, however, especially those in the Earth's oceans, are best modelled by two or more layers of fluid. The use of multiple layers more accurately represents the vertical stratification of the ocean whose surface layers are less dense than its abyss. For example, Agulhas eddies have well-defined surface signatures and it may, at first thought, be expected that they are surface-trapped features and are adequately modelled as single-layer 'blobs'. However, it is interesting to note that satellite observations show that some Agulhas eddies are influenced by the Walvis Ridge (a steep sea-floor topographic feature). Some eddies are able to get over the ridge but some are trapped by the ridge. This suggests that there is a deep lower layer flow associated with Agulhas eddies and that it is appropriate for them to be modelled by two (at least) layers.

Even the problem of the interaction of single-layer, *non*-quasi-geostrophic vortices with topography raises interesting questions. For instance, large Rossby number vortices (i.e. those for which planetary rotation is relatively unimportant), such as those occurring in the surf-zone, are known to move along contours of constant depth (or, more generally, constant potential vorticity) in opposite directions according to the sign of their vorticity. On the other hand, small Rossby number vortices also move along contours of constant depth but in the *same* 'westward' direction, irrespective of the sign of their vorticity. It would be an interesting study to examine the effect of varying the Rossby number from very small to very large values to see its effect on, for example, the direction and speed of motion. Is there, for example, a critical Rossby number and form of topography such that the vortex is stationary?

Another aspect of vortex motion that has received little attention in the literature is multiple vortex interaction. This is relevant to Agulhas eddies and Gulf Stream rings, where vortices do not occur in isolation but rather in groups. In fact, Agulhas eddies tend to form a train travelling northwestward into the Atlantic. How does such an arrangement of vortices affect speed and rate of decay? For example, are Agulhas eddies arranged in such a manner to minimise their Rossby wave drag? It would perhaps be fortuitous if they were. How do their collective wakes interact? Both mathematical and numerical techniques are presently being used to shed light on such questions.

The interaction of vortices with pre-existing waves, especially Rossby waves, rather than the generation of waves by moving vortices is a topic that has received little attention. This is known as the scattering problem and is a well-developed topic in applied mathematics, but has not hitherto been applied to vortices and Rossby waves depite the fact that such waves are ubiquitous in geophysical fluids and, inevitably, vortices will encounter them. Can the life of a vortex be prolonged by the absorption of energy from the scattering of Rossby waves? Perhaps this explains the longevity of Jupiter's Great Red Spot, with its apparent resistance to energy loss by wave radiation. Such a study could, in the first instance, be done using a mathematical model based on quasigeostrophic theory.

Big advances in satellite observation technology and data processing techniques (such as removal of atmospheric effects, improved calibration and signal to noise performance) leading, in particular, to their ability to resolve mesoscale ocean features such as vortices, are keenly anticipated within the next few years. Coupled with improving numerical techniques and the greater understanding of simple models through analytical work, the near future promises to be an exciting time in the study of the motion of geophysical and astrophysical vortices.

## 9.5 Acknowledgements

I am grateful to David Dritschel, GertJan van Heijst, Luis Zavala Sanson and David Dunn for supplying figures for this article.

## 9.6 Further reading

Dowling, T. E. 1995 Dynamics of Jovian atmospheres. *Ann. Rev. Fluid Mech.* **27**, 293–334.

Flierl, G. R. 1987 Isolated eddy models in geophysics. *Ann. Rev. Fluid Mech.* **19**, 493–530.

McDonald, N. R. 1999 The motion of geophysical vortices. *Phil. Trans. Roy. Soc. Lond.* A **357**, 3427–3444.

McWilliams, J. C. 1985 Submesoscale, coherent vortices in the ocean. *Rev. Geophys.* **23**, 165–182.

Nezlin, M. V. & Sutyrin, G. G. 1994 Problems of simulation of large, long-lived vortices in the atmospheres of the giant planets (Jupiter, Saturn, Neptune). *Surv. Geophys.* **15**, 63–99.

van Heijst, G. J. F. 1994 Topography effects on vortices in a rotating fluid. *Meccanica* **29**, 431–451.

# 10
# Earth's future climate

## Mark A. Saunders

*Department of Space and Climate Physics, Benfield Greig Hazard Research Centre, University College London, Holmbury St Mary, Dorking, Surrey RH5 6NT, UK*

Climate and weather affect us all in our daily lives, impact the performance of much of industry, and lead to tens of billions of pounds of damage worldwide each year. Governments and the general public alike are becoming concerned as press reports, personal experience and anecdotal information all point to an increase in the frequency and severity of extreme weather events linked to climate change. Many opinions have been expressed on climate change from the doomladen to the dismissive. This chapter aims to state clearly the current scientific position on climate change, its impacts, and the effects we may expect in this millennium. This will help decision makers, managers of weather risk, and all with an interest in our future climate.

## 10.1 Introduction

Let us begin by defining the meaning of 'climate change'. The United Nations Framework Convention on Climate Change uses the term to describe changes in the average state of the weather brought about only by human (anthropogenic) activities. These include in particular those processes which emit the heat-trapping gases carbon dioxide and methane into the air. A more general definition, common in the scientific community, refers to change brought about by any source, human as well as natural. Here we use 'climate change' in its widest sense to describe any change in the Earth's climate on timescales longer than a few months. Thus we consider (i) interannual (year-to-year) climatic changes linked to natural variability caused, for example, by El Niño and volcanic eruptions, as well as (ii) multidecadal trends linked to anthropogenic global warming or to changes in solar output. On the interannual to decadal level, natural climatic variability has a far greater impact on local climate than long-term trends due to global warming. However, on the multidecadal (≥ 50 years) level, changes in the mean climate due to global warming could, in certain regions, start to approach the limits of current natural variability.

Are weather-related disasters becoming more common as the news headlines often indicate? We examine this by considering the statistics on natural catastrophes reported by insurers and reinsurers. For the recent 10-year period (1990–1999), the economic (i.e. total) and insured losses due to natural catastrophes have averaged £34 billion and £7 billion per year respectively. Windstorms alone (hurricanes, winter storms and tornadoes) account for 30 per cent of these economic losses and for 70 per cent of the insured losses. In 1998, for example, the devastation brought by Hurricane Georges in the Caribbean and the US (see Figure 10.1) proved the largest insured loss of the year and the third most costly in American insurance history. Since 1970, the number and cost of natural catastrophes have risen continuously. Comparing 1990–1999 with the decade 1970–1979, shows that the number of major natural catastrophes, and their resulting economic losses and insured losses, have increased by factors of 1.8, 4.3 and 9.2 respectively (figures inflation adjusted, Munich Re.). The largest rises have occurred since the late 1980s. Most of the loss increases may be explained by socioeconomic factors such as a higher density of population in hazard-prone areas, and by more assets being insured in hazard areas. However, it seems unlikely that the near two-fold increase in the number of major natural disasters can be due merely to improved reporting. Many insurers feel that the frequency of extreme events has genuinely increased and that long-term climate change could be a contributory factor.

A number of informative and influential summaries on the effects of anthropogenic global warming have appeared in recent years (see 'Further reading'). This review builds on and extends these summaries by treating climate change in its broadest sense (i.e. considering the impact of natural variability in addition to long-term global warming trends), and by including new research on hurricane trends. A more comprehensive and technical account of this review may be found in the Millennium Issue of *Philosophical Transactions*.

**Figure 10.1.** Residents in Key West, Florida, fleeing Hurricane Georges on 26 September 1998. Georges claimed the lives of 500 people across the Caribbean, caused economic damages of £5 billion (second highest natural catastrophe loss of 1998) and insured losses of £2.5 billion (largest of 1998). Georges also contributed to making the 1995–1999 five-year Atlantic hurricane total of 41 the highest such total on record. (Image used with permission of Associated Press.)

## 10.2 Types and causes of climate change

For brevity we consider two types of climate change: those producing inter-annual change, and those producing multidecadal trends. Year-to-year changes in climate, especially those linked to changing temperature, rain-fall and windspeed impact much of industry. It is also important to recog-nise and plan for long-term trends. Although these perforce allow governments and industry more time to respond, thereby minimising human and financial impacts, reaction times are often long. Furthermore, slow multidecadal climate trends may also affect interannual variability, as suggested recently with regard to El Niño frequency.

## 10.2.1 Interannual changes

The three most important classes of natural variability of the Earth's climate are those associated with El Niño/La Niña, the North Atlantic Oscillation, and major volcanic eruptions.

El Niño and its cold-episode sister La Niña are the strongest interannual climate signals on the planet. Global damage estimates for the 1997/98 major El Niño event exceed £20 billion. Archaeological evidence suggests that El Niños and La Niñas have been occurring for at least 15 000 years. The clearest sign that an El Niño (La Niña) event is underway is the appearance of unusually warm (cold) water between the Date Line and the coasts of Ecuador and Peru. During the 1997/98 event, for example, waters in this region were about 5 °C warmer than usual. However, El Niño (La Niña) is more than just a warming (cooling) of the eastern tropical Pacific – it is a perturbation of the ocean–atmosphere system; hence it is also called ENSO (El Niño Southern Oscillation), where the Southern Oscillation refers to the accompanying large-scale seesaw oscillation in atmospheric pressure between the Pacific and Indian Oceans. A major El Niño or La Niña event occurs about every four years and leads to abnormal patterns of temperature, rainfall and storminess around the globe. The amplitudes of many of these climate anomalies currently exceed the mean changes likely by 2100 due to anthropogenic global warming.

The North Atlantic Oscillation is the major source of interannual variability in the atmospheric circulation over the North Atlantic and surrounding land masses. It is defined in terms of the pressure difference between Iceland and the Azores. Year-to-year and decadal changes in the North Atlantic Oscillation are linked to seasonal changes in regional temperature, rainfall and storm occurrence over Europe and, to a lesser degree, over north-eastern Canada and the eastern USA. Unpublished research by the author, for example, shows the North Atlantic Oscillation is linked directly to 50 per cent of the year-to-year variability in UK winter storminess, to 50 per cent of the winter rainfall in Scotland, Spain and parts of Norway, and to 40 per cent of the winter temperature anomalies over much of western Europe. The oscillations in local winter climate linked to the North Atlantic Oscillation currently exceed the projected changes in mean climate expected from global warming by 2100.

Volcanic eruptions are also an important source of natural climate variability. Major explosive eruptions can inject dust and sulphate aerosols (microscopic solid particles of diameter $10^{-3}$m to $10^{-6}$m) to heights of 20 km in the stratosphere. These particles reflect solar radiation and have a lifetime of over a year. The net effect is a cooling of the Earth's surface. The eruption of Mount Pinatubo in the Philippines in June 1991 stands out from a climatic point of view as one of the most important eruptions this century. During 1992, global surface temperatures cooled by 0.3 to 0.5 °C due to the eruption. The eruptions of Krakatoa (1883) and Tambora (1815) also led to global climate cooling. Indeed, 1816 has been called the 'year without a summer' when crops failed in Europe and the USA.

## 10.2.2  Multidecadal trends

The two most important factors influencing future long-term trends in the Earth's climate on the 10–1000 year timescale are anthropogenic global warming (the enhanced greenhouse effect) and changing solar output. The effect of changes in the Earth's distance from the Sun and changes in the tilt of the Earth's poles towards the Sun – which have periods of about 100 000, 41 000 and 22 000 years and are believed to be a major contributory factor to ice ages – will not be significant on this time frame.

The basic cause of global warming is described in terms of the enhanced greenhouse effect (Figure 10.2). The Earth has a natural greenhouse effect which keeps the planet's surface 33 °C warmer than it would otherwise be; at an average temperature of 15 °C rather than −18 °C. This greenhouse warming is due to atmospheric gases (called greenhouse gases) that trap parts of the Earth's surface infrared heat which is trying to escape into outer space. The main natural greenhouse gases are water vapour, carbon dioxide, methane, and nitrous oxide. The enhanced greenhouse effect comes from an increase in the concentration of these natural greenhouse gases due to human activities. This increase has been taking place since the start of the Industrial Revolution in c. 1765. The order of importance in contributing to human-induced global warming is carbon dioxide (70 per cent), methane (20 per cent) and nitrous oxide (10 per cent). Quantities of these greenhouse gases are increasing steadily in the atmosphere due to fossil fuel burning (coal, oil and natural gas), deforestation, and rice cultivation. The use and production of energy accounts for about 60 per cent of global greenhouse emissions, the burning of forests contributes about 10 per cent, and rice fields and decaying rubbish a further 10 per cent.

The main reason for carbon dioxide having the greatest potency of all

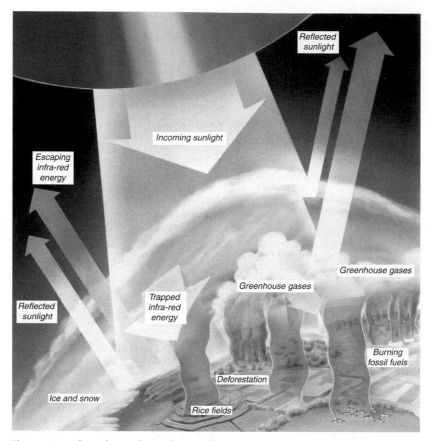

**Figure 10.2.** The enhanced greenhouse effect. The Earth's surface temperature is slowly rising due to human activities which are releasing heat-trapping gases, notably carbon dioxide and methane, into the atmosphere. By 2100 this temperature rise is expected to reach $2 \pm 1\,^{\circ}\mathrm{C}$, other climatic influences remaining constant. This will be the fastest rate of climate change the Earth has experienced since the start of modern civilisation 10 000 years ago. (Figure from *Global Warming* by Pringle ©1988, Discover Publications. Used with permission of Hodder and Stoughton Limited.)

the human-caused greenhouse gases is its persistence: its lifetime in the atmosphere is about 100 years. Carbon dioxide atmospheric concentrations have increased since the pre-industrial period from about 280 ppmv (parts per million by volume) to about 360 ppmv in 1997. We know this from analysis of ice cores and, since the late 1950s, from precise, direct measurements of atmospheric concentration. That the observed increase in atmospheric carbon dioxide comes from anthropogenic activity is evident from the close agreement between the long-term rise in atmospheric carbon dioxide and the increase in carbon dioxide emissions. Such emissions now stand at over 6000 million tonnes annually.

Temperatures have not increased as much as one would expect from the observed carbon dioxide increase. The reason for this is thought to be the mitigating effect of industrial aerosols, especially sulphate aerosols. These differ from the aerosols we are familiar with in our daily lives, such as in hairsprays. They are microscopic solid particles injected into the atmosphere by either natural events such as dust storms and volcanic eruptions, or by anthropogenic activities such as fossil fuel and biomass burning, and changing land use. By blocking incoming solar energy, these aerosols act to cool surface temperatures and thus mitigate global warming. Research by the Hadley Centre for Climate Prediction (Bracknell, UK) and other modelling groups shows that aerosols will not globally cancel global warming but could offset locally a significant amount. Despite this, uncertainties remain in estimating the global and regional climatic impact of aerosols.

Changes in the Sun's radiative output also cause long-term climate change. Many climatologists support the view that variations in solar activity have produced significant changes in the Earth's climate during the last millennium. For example, in the eleventh and twelfth centuries, when solar activity (i.e. sunspot numbers) was high, the Earth was significantly warmer. During this period the Vikings inhabited Greenland. As solar activity waned, so did the Vikings' fortunes, and by the cold fourteenth century they were struggling to survive. The late seventeenth century witnessed another inactive period on the Sun – the Maunder Minimum – when sunspots virtually disappeared. This period coincided with the 'Little Ice Age' when alpine glaciers expanded, the River Thames froze regularly in winter, and temperatures dropped significantly.

Despite the above examples, the Sun's role in climate change remains poorly understood. What is certain, however, is that the Sun has contributed little to the 0.4 °C rise in global surface temperature which has occurred since the mid 1970s. Satellites have been directly monitoring solar radiative output over this period. These records show that solar output has remained almost constant. This result is evident in Figure 10.3 which shows reconstructed solar and climate time series dating back to

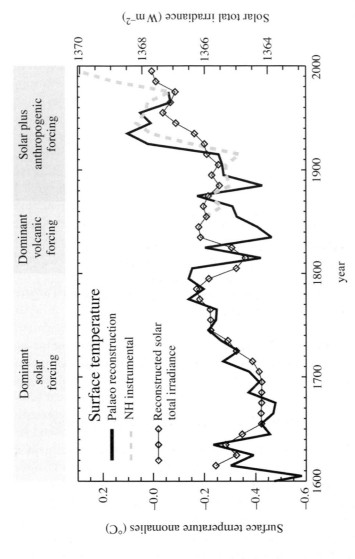

**Figure 10.3.** Comparison for 1600–1995 of the reconstructed solar total irradiance with the reconstructed northern hemisphere surface temperature record. Prior to 1860, the latter is based largely on tree ring growth. Changing solar radiation can explain most of the long-term changes in temperature prior to the industrial revolution but can not account for the rise in temperature since the 1970s. (Figure courtesy of Judith Lean.)

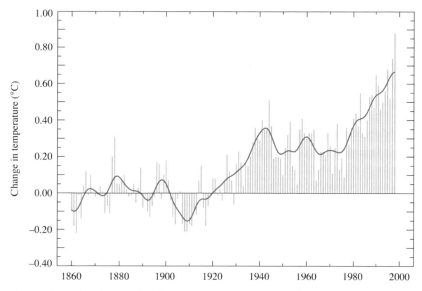

**Figure 10.4.** Global annual surface temperature variations from 1860 to 1998. Values are shown as departures (°C) from the mean at the end of the nineteenth century. The smoothed curve is obtained using a 21-year filter. Mean global temperatures are now higher than at any time since 1400. (Figure courtesy of the Hadley Centre for Climate Prediction and Research, Bracknell, UK.)

1600. While these series correlate strongly in earlier epochs, in particular during the period 1610–1800 which includes the Maunder Minimum, this is not the case since 1970.

## 10.3  Evidence for twentieth century global climate change

This section describes the changes in global temperature, precipitation, sea level and weather extremes observed during the twentieth century.

Reliable global temperature records extend back to 1860. Figure 10.4 shows the change in global mean surface temperature from 1860 through to 1998. This temperature has increased by 0.6 to 0.7 °C since the late nineteenth century, and by 0.4 °C over the last 25 to 30 years, the period with most reliable data. The warming has occurred largely during two periods, between 1910 and 1940, and since the mid 1970s. 1998 was the warmest year ever (followed in order by 1997, 1995, 1990 and 1999), and 15 of the 16 warmest years on record have now occurred since 1980. The warming trend has not been uniform; for example, the continents between 40°N and 70°N in winter and spring have warmed most, and a few areas such as the North Atlantic Ocean have even cooled. Various indirect evidence supports the recent warming of global temperature. For example, most glaciers worldwide are shrinking. European alpine glaciers have lost over one-third of their surface area and well over half of their volume in the last 100 years. This retreat is consistent with a warming of 0.6 to 1.0 °C.

In order to establish whether the twentieth century warming is caused by anthropogenic factors or is part of the natural variability of the climate system, it is helpful to examine climate changes which have occurred in the past. The available evidence indicates that recent decades have been the warmest in the last 600 years and that the Earth has only been warmer than now for less than 10 per cent of its geological lifetime. A further warming of 1 to 2 °C would make the Earth the warmest it has been in 150 000 years.

A sceptic may wonder how much reliance can be placed on ancient historical data. Confidence that the deduced temperature variations are indeed real comes from indirect information such as the painting in Figure 10.5 showing skating on Dutch canals. The picture was made in the early seventeenth century during the Little Ice Age when global temperatures were about a degree colder than they are now. It shows that winter freezes were characteristic of the time, with skating and frost fairs regular features on European canals and rivers. Such events are, of course, now rare. The winter of 1894–95 was the last occasion, for example, when the River Thames froze sufficiently to allow an ice fair.

Surface warming increases evaporation and the amount of water vapour in the atmosphere. There is thus greater potential for increased global rainfall. Analyses of precipitation changes during the twentieth century reveal that rain plus snowfall has increased by 2 per cent globally (equivalent to an extra 22 mm of annual rainfall everywhere). However, not all parts of the world have seen a precipitation increase since 1900. In the African Sahel and Indonesia, for example, rainfall has decreased.

Over the last 100 years global sea level has risen by between 10 and 25 cm based on analyses of tide gauge records and allowing for uncertainty from vertical land movements. Much of this rise is due to the increase in global temperature since 1900. The warming and consequent thermal expansion of the oceans accounts for 2 to 7 cm of the observed rise, while the retreat of mountain glaciers and ice caps accounts for a further 2 to 5

**Figure 10.5.** Winter picture painted in 1601 by Peter Brueghel the Younger, showing skaters on the frozen canals of Holland. The cold winters necessary for the Dutch canals to freeze were regular features of this time (the Little Ice Age). Skating on these canals has now become a rare event due to rising global temperatures. (Figure courtesy of the Kunsthistorisches Museum, Vienna, Austria. Erich Lessing/Art Resource.)

cm. Other factors such as the contribution from the huge ice sheets of Greenland and Antarctica are more uncertain.

Is there evidence for trends in the global and/or regional number and intensity of hurricanes and floods in recent decades? New research by the author and F. P. Roberts, using the latest available data, show that for the period 1969–1999 significant multidecadal variability exists in the annual number of tropical cyclones (TCs) of various intensities for the northern hemisphere as a whole. While TC numbers rose from the mid 1970s to a peak in the early 1990s they are no more frequent now than at the beginning of the period (Figure 10.6). Furthermore, interannual variability of TC activity is negatively correlated with global surface temperature. We conclude that, contrary to popular belief, global warming has not had a significant influence on northern hemisphere TC numbers during the 1969–1999 period.

Further recent research shows evidence for extratropical windstorms

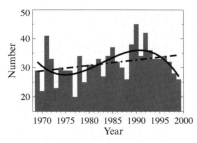 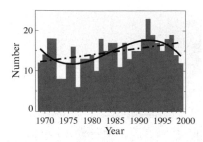

**Figure 10.6.** Time series for 1969–1999 of the number of hurricane-strength (maximum sustained wind >33 m s$^{-1}$) (left panel), and intense hurricane-strength (maximum sustained wind >50 m s$^{-1}$) (right panel), tropical cyclones for the three main northern hemisphere basins combined. The linear trends (dashed line) are upward but the cubic trends (solid lines), which provide the best overall fits, show that northern hemisphere hurricane-strength and intense hurricane-strength events have been falling in number since the early 1990s. With the 1990s containing the seven warmest years on record, this suggests that global warming is not increasing the number of hurricane-strength events as popularly believed. (Figure courtesy of Frank Roberts and Mark Saunders.)

in the North Atlantic having increased in intensity in recent decades (Figure 10.7). During the late 1980s and early 1990s, winter windspeeds over the UK and northwest Europe reached their highest levels since the 1920s. It remains to be determined whether this rise is related to anthropogenic global warming or is due to natural variability of the climate system.

In addition to the increase in total precipitation since 1900 noted above, various studies support an upward trend in the proportion of the total precipitation occurring in the extreme precipitation category (i.e. >50 mm per day). In the USA, for example, this percentage has increased from less than 8 per cent in 1900 to more than 10 per cent now. This trend has been been linked to the recent rise in the number of USA floods, a notable example being the devastating Mississippi flood in the summer of 1993. One might expect that susceptibility to flooding and flash floods in other countries may also be rising, but research results on this are not yet available.

**Figure 10.7.** Porthleven, Cornwall, UK, being battered by 100 m.p.h. winds on 4 January 1998. The Christmas and New Year 1997/98 storms caused damage of £1.6 billion across Europe. New research indicates that the 1990s were more stormy for northwest Europe than any of the three previous decades. Whether this trend is due to climate change or to multidecadal natural variability remains unclear. (Image used with permission of Apex Photo Agency Ltd.)

## 10.4 Predicted future climate change

This section describes the 'best estimate' projections for how the Earth's climate will change due to global warming by 2100. Forecasts cannot be made with any certainty beyond this time. These predictions are made by sophisticated climate models which describe mathematically and physically the different coupled parts of the climate system; namely the atmosphere, ocean, land and ice sheets. Although much remains to be done to narrow the uncertainty of model predictions, there are sound grounds for believing that current climate models are able to simulate important aspects of anthropogenic climate change. This confidence comes from the success of these models in simulating aspects of historical climate change.

If the concentration of atmospheric greenhouse gases continues to rise at the current 0.7 per cent annual rate, the 'best estimate' consensus of the major climate models is that global mean surface temperatures will rise by about 2 °C between 1990 and 2100. This projection takes into account the effects of future changes in aerosols and the delaying effects of the oceans. The oceanic inertia means that the Earth's surface and lower atmosphere would continue to warm by a further 1–2 °C even if greenhouse gas concentrations stopped rising in 2100. The range of uncertainty in the global warming projection to 2100 is 1 °C to 3.5 °C. A 1 °C rise would be larger than any century-scale trend in the past 10 000 years. Another consequence of this warming would be an increased probability of heat waves. For example, in central England by 2050, summers as hot as 1995 would occur every 3 years instead of once every 50 years as seen during the twentieth century.

Enhanced evaporation will lead to a global increase in precipitation of between 3 and 15 per cent by 2100. Although trends at local levels are less certain, some areas such as southern Europe in summer and Australia are expected to see a decrease in precipitation. A high spatial resolution model prediction for the UK and northwest France is shown in Figure 10.8. Mean winter precipitation is shown for the 2020s and 2050s expressed as a percentage change from the 1961–1990 average. Winters are significantly wetter throughout the UK, and by up to 10 per cent in southeast England by the 2050s.

The Earth's average sea level is predicted to rise about 50 cm by 2100. The uncertainty range is large – 15 to 95 cm – and changing ocean currents could cause regional sea levels to rise much more or much less than the global average. About 70 per cent of this rise will come from thermal expansion of the upper layers of the ocean as they warm. Melting alpine glaciers will contribute a further 20 per cent of the rise. The Greenland and Antarctic ice sheets are not expected to contribute significantly to sea level rise as their melting will be balanced by increased snowfall in both regions. As the warming penetrates deeper into the oceans, sea level will continue rising well after surface temperatures have levelled off.

The future prospects concerning windstorms are unclear. Current climate models, of grid size about two degrees, lack the spatial resolution to properly simulate tropical storms and hurricanes and are thus unable to make worthwhile long-range projections. However, the author's results presented in Section 10.3 indicate that to date global warming has not significantly influenced northern hemisphere tropical cyclone frequency. Current thinking suggests an increase in the frequency and severity of

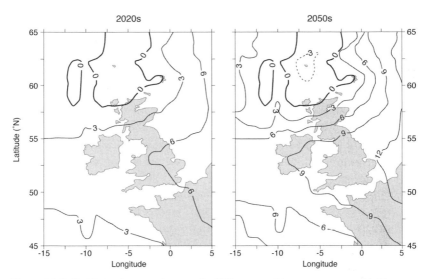

**Figure 10.8**. Precipitation trends over the UK expected in a warmer world. The figure shows the percentage change in mean winter precipitation for the 2020s (left) and 2050s (right) decades with respect to the average of 1961–1990. (Figure reproduced courtesy of Climate Change Impacts UK (1996); Crown copyright is reproduced with the permission of the Controller of Her Majesty's Stationery Office.)

European windstorms as winters become progressively milder. Models point to windspeed return periods shortening over the northern UK in winter, with a daily mean windspeed of $18\,\mathrm{m\,s}^{-1}$ changing from a one-in-two year event to an annual event by 2050. Over the southern UK, little change in return periods is expected for the strongest winds.

The future prospects with regard to flooding are clearer. Climate models generally agree on an increasing tendency for more intense but less frequent precipitation in a warmer world. In many areas, the frequency of days with precipitation of high intensity will increase at the expense of days with precipitation of low intensity. This shift in precipitation intensity will lead to more floods, and arguably to more massive storms and hailstorms.

Can we change the course of climate change? While the natural variability associated with El Niño/La Niña and other interannual climate changes cannot be altered realistically at present, we can alter the course of anthropogenic global warming. The cuts in greenhouse gas emissions agreed by industrialised countries at the United Nations Convention on Climate Change in Kyoto in December 1997 – to levels 5.2 per cent below 1990 concentrations by 2008–12 – though welcome, will have little impact on slowing climate change. Global reductions in carbon dioxide emissions of around 60 per cent are necessary to prevent greenhouse gas concentrations from rising still further.

## 10.5  Future challenges for climate science

This section offers speculation for long-term climate change beyond 2100, and reviews the major challenges now facing climate science.

Projections for climate change beyond 2100 are speculative and subject of course to political decisions on greenhouse gas emissions. However, one can expect that our long-range skill in predicting ENSO and its impacts, landfalling hurricanes, European winter storms, and seasonal extremes in temperature and rainfall, will all improve steadily. Business, industry, and commerce will be able to plan ahead in the knowledge that unexpected climate events are unlikely to impair performance. It is even possible that we will learn how to actively modify and control regional climate to our benefit thereby minimising the impacts of droughts, floods, windstorms and heatwaves.

It is also certain that surprises will happen. Major volcanic eruptions, such as Pinatubo or Krakatoa, will continue to occur giving rise to global impacts on climate lasting a year or two. Based on the last 1000 years, we can expect that the Sun's radiation output will continue to vary, probably causing global temperature changes of up to 1 °C during the present millennium (cf. the Maunder Minimum and Little Ice Age of the seventeenth century). Should an asteroid or small comet collide with the Earth, as has happened in the recent geological past, the impacts on climate could be even greater.

Prior to the early 1980s there was little scientific and public interest in climate change. The significance of El Niño and its impacts was unappreciated, La Niña was unknown, and the role of the North Atlantic Oscillation and other teleconnection patterns on interannual climate variability was unrecognised. This situation has now changed completely. Today climate change is constantly in the news, major research programmes are underway to improve climate prediction, and enterprising industries are recognising the impact of climate and weather on their business performance. Despite these advances, uncertainties abound at almost all levels of climate forecasting. These uncertainties have even led some scientists to question how much of the current warming can be attributed to human activities. Considerable research effort is underway to reduce these uncertainties.

Arguably the greatest challenges now facing climate science are to reduce uncertainties in: (1) the effect of aerosols on climate change; (2) the role played by clouds; (3) the effect of climate change on 'flipping systems' such as the North Atlantic circulation pattern called the 'Atlantic Conveyor Belt'; and (4) how the frequency and intensity of extreme weather events, especially windstorms, will change in a warmer world.

Aerosols, especially sulphate aerosols, promote climate cooling by reflecting solar radiation back into space (Section 10.2.2). However, calculations for the effect of aerosols on climate change through to 2100 are very uncertain due to several factors. First, scenarios for how sulphur dioxide emissions will change are uncertain. Recent estimates suggest that future emission rates will be lower than envisaged originally. Second, recent models generate a lower sulphate aerosol concentration per tonne of sulphur dioxide emitted. Third, sulphate aerosols can also cool climate by changing the reflectivity and longevity of clouds. This indirect effect now appears as important as the aerosol direct effect but, to date, has not been included in climate models. Fourth, the significance of other types of aerosol such as carbon and soot (which both warm the atmosphere) requires proper assessment.

Clouds can both warm and cool the Earth. They prevent heat from the Earth's surface escaping into space, thereby warming the Earth, and also reflect solar radiation back into space, thereby cooling the Earth. The net effect can be positive or negative depending on the height, temperature and reflecting properties of the clouds, all of which vary in time and from place to place. Our current understanding of the effect of clouds on climate change is poor and this forms one of the largest uncertainties.

Few climate systems behave linearly where doubling the input doubles the output. A feature of such complex systems is that small changes in the forcing conditions (i.e. the concentration of greenhouse gases) could lead to abrupt changes or even to 'flipping' occurring. Some climate models suggest that circulation patterns in the North Atlantic Ocean could be susceptible to this type of disruption if more fresh water entered the Arctic Ocean as a result of global warming. Should this happen, the Gulf Stream, which is responsible for half the heat received by the UK and northwest Europe during winter, would move equatorward, leading to a cooler Europe while the rest of the world warmed. The reason for the range of responses in climate models on this issue is not fully understood. However, happily, Atlantic circulation patterns and the Gulf Stream remain steady to date.

If changes in the North Atlantic circulation did occur, the resulting shifts to patterns of sea surface warming and cooling would affect European winter storminess. Even without this complication, the impact of global warming on the frequency and intensity of extreme weather events, especially hurricanes and windstorms, is the least understood potential impact of climate change. Further research is required on this and on the equally important issue of improved seasonal forecasting of land-falling hurricanes and winter storms.

In conclusion, we are witnessing a unique and exciting era in climate research. The pace of progress and discovery these past 15–20 years will surely continue. As advances come in computing power, in the physics of climate models, and in the quality and length of the historical climate record, so will our ability to better forecast climate change and its impacts . . . this to the benefit of all humankind.

## 10.6  Further reading

Hulme, M. & Jenkins, G. J. 1998 *Climate change scenarios for the UK: scientific report*, UKCIP Technical Report No.1, Climatic Research Unit, Norwich, 80pp.

Houghton, J. T. 1997 *Global warming: the complete briefing*, 2nd edn. Cambridge University Press, 251 pp.

IPCC 1996 Summary for policy makers. In *Climate change 1995 – the science of climate change. The Second Assessment Report of the Intergovernmental Panel on Climate Change: Contribution of Working Group I*. J. T. Houghton et al. (eds.), Cambridge University Press, 572 pp.

Karl, T. R., Nicholls, N. & Gregory, J. 1997 The coming climate, *Sci. Am.* May pp. 78–83.

Saunders, M. A. 1999 Earth's future climate, *Phil. Trans. R. Soc. Lond.* A, **357**, 3459–3480.

# Contributor biographies

## J. M. T. Thompson

Born in Cottingham, Yorkshire, on 7 June 1937, Michael Thompson studied at Cambridge, where he graduated with first class honours in Mechanical Sciences in 1958, and obtained his PhD in 1962, his ScD in 1977. He was a Fulbright researcher in aeronautics at Stanford University, and joined University College London (UCL) in 1964. He has published four books on instabilities, bifurcations, catastrophe theory and chaos, and was appointed Professor at UCL in 1977. Michael was elected a Fellow of the Royal Society in 1985, and was awarded the Ewing Medal of the Institution of Civil Engineers. He was a senior SERC Fellow and served on the IMA Council. In 1991 he was appointed Director of the Centre for Nonlinear Dynamics at UCL. He is currently Editor of the Royal Society's *Philosophical Transactions* (Series A) which is the world's longest running scientific journal. Scientific interests include nonlinear dynamics and their applications: recreations include walking, tennis and astronomy with his grandson Ben, shown below.

## J. Magueijo

João Magueijo was born in Évora, Portugal, in 1967. He studied physics at Lisbon University, perhaps the reason for his swift move to Britain in 1989. He got his PhD at Cambridge in 1993, following which he became a Research Fellow at St. John's College, Cambridge. Since then his interests have oscillated between observational and lunatic aspects of cosmology. The former concern data analysis issues associated with the cosmic radiation. The latter include work on topological defects, super-symmetric inflation, cosmic magnetic fields, etc. Since 1996 he has been at Imperial College, first as a Royal Society University Research Fellow, then as a Lecturer. His most recent work includes an attempt to build cosmology upon theories permitting the variation of the fundamental constants of nature. Recreations include annoying his co-author Kim Baskerville.

## K. Baskerville (top right)

Kim Baskerville was born in Melbourne, Australia in 1968, where she remained just long enough to complete a BSc (Hons) degree in physics at Melbourne University. She then exchanged her sunny native shores for the damp fog of Cambridge, where she completed a PhD. She then held post-doctoral positions first at the University of Wales in Swansea, then at Durham University. However, she found the chauvinistic atmosphere of British universities uncongenial, and in 1999 left research to begin a new career teaching physics in high school. When not marking books, she enjoys reading and foreign travel – to anywhere warmer than Britain.

## J. García-Bellido

Born in 1966 in Madrid, Spain, Juan García-Bellido studied in the Autónoma University where he graduated with honours in theoretical physics in 1988 and obtained his PhD in 1992. He was a Postdoctoral Fellow at Stanford University (1992–94), University of Sussex (1995–96), Fellow of TH-Division CERN (1996–98), Royal Society University Research Fellow at Imperial College (1998–99). He is at present Lecturer at the Autónoma University of Madrid. He has published around 50 papers in theoretical physics; has attended, organised and given lectures at international conferences and summer schools around the world; and is referee of the most prestigious journals in the field. He is married to another theoretical physicist and has a young daughter. Scientific interests include the early universe, black holes and quantum gravity. Non-scientific interests include classical music and hand drawing.

## B. Moore

Ben Moore was born in Falstone, Northumberland in 1966. He received a first class honours degree in astronomy and astrophysics from Newcastle University in 1988. After obtaining his PhD from Durham University in 1991 he spent several years as a research associate at the University of California, Berkeley and at the University of Washington, Seattle. He returned to the UK in 1996 to take up a Royal Society University Research Fellowship at Durham University. His scientific interests within the field of cosmology include galaxy formation and the origin of structure in the universe: recreational activities include rock-climbing, mountaineering and snow-boarding.

## M. D. Gray

Born in Barnstaple, Devon, in 1962 Malcolm Gray read physics with astrophysics at Birmingham, graduating with first class honours in 1983. He obtained his D.Phil. from Sussex University in 1987. After four years post-doctoral research experience in David Field's group at the University of Bristol, and a brief period working as a physicist for the British Antarctic Survey in Cambridge, he was awarded a Royal Society University Research Fellowship in 1991. He initially held the fellowship at Bristol (first in the School of Chemistry, then in the Department of Physics) before moving to Cardiff University in 1998. Originally a theoretical/computational astrophysicist, his work has broadened to cover observations and instrumentation. Scientific interests, in addition to masers, include more general radiation transfer, cosmic dust and gas-phase elemental depletions and rate-coefficients for hydrogen on small astrophysical molecules. Recreations include skiing, flying and hill-walking.

## A. J. Coates

Andrew Coates, born at Heswall, Cheshire, in 1957, gained a BSc (first) in physics from UMIST in 1978, and MSc (1979) and D.Phil. (1982) in plasma physics from Oxford University. He has been at UCL's Mullard Space Science Laboratory since. He was a Royal Society University Research Fellow and is now Reader in Physics. He was guest scientist at MPAe, Germany and University of Delaware, and was a media fellow at BBC World Service. Space mission involvements include AMPTE, Giotto, Meteosat, Cluster and Cassini where he led the team providing the electron spectrometer (shown in photograph). He is on several PPARC committees and the ESA Solar System Working Group. Scientific interests include the solar wind interaction with planets and comets and space instrumentation; he has over 100 publications. Interests include popularising science, skiing and helping bring up twin daughters.

## S. K. Dunkin

Born in 1973, Sarah went to Walworth School in South London. Fascinated with astronomy and the Moon from a very early age, Sarah studied astronomy at University College London and obtained a first class honours degree in astrophysics in 1994. Staying at UCL, she completed her PhD three years later, studying the spectroscopic properties of Vega-like stars and the dust around them. After completing a three-year PPARC Postdoctoral Research Fellowship, she now holds a Royal Society Dorothy Hodgkin Fellowship at UCL, and works at the Rutherford Appleton Laboratory. Sarah especially enjoys communicating science to the younger generation and the general public. Her hopes are that everyone will have the opportunity of travelling to the Moon, perhaps not in her lifetime, but in the not too distant future.

## D. J. Heather

David Heather was born in 1973 and is currently a European Space Agency Research Fellow at the European Science and Technology Research Centre in the Netherlands. He completed his PhD research in the Planetary Geology Group at University College London, using Clementine data to analyse fresh impact craters on the lunar surface, and obtained his Bachelor's degree in astronomy from the University of Hertfordshire in 1995. David is very keen to advance the communication of science to the general public at all levels, from the enthusiastic amateur to the young 'scientists to be'. In this capacity, he has written and taught several courses in astronomy for evening classes and school groups. By making science accessible to the public, especially to the younger generation, David hopes he will be encouraging some of those who will one day explore the surfaces of the planets to follow their dreams.

## L. Vočadlo

Lidunka Vočadlo studied at University College London, graduating with upper second class honours in physics and astronomy in 1988. After gaining a PGCE at the Institute of Education in 1989, she obtained her PhD in 1993 in the Geological Sciences Department. She has remained there ever since, doing post-doctoral research from 1993 to 1996 and as a UCL Research Fellow from 1996 to 1999. In 1998, Lidunka was awarded the Dornbos Memorial Prize for young researchers, and in 1999, she was awarded a Royal Society University Research Fellowship. Her research interest lies in the computer simulation of Earth and planetary materials; she relaxes with long sunny days on the fells in the Lake District and long cool evenings drinking fine wine, watching them slide into the night.

## D. Dobson

After graduating from Bristol University in 1991 with upper second class honours in geology, David Dobson moved to University College London in 1992 to study for a PhD. David obtained his PhD in 1995, has since been a post-doctoral research fellow and is currently a NERC fellow, both at UCL. While at Bristol, David was awarded the Donald Ashby Prize and he was the 1999 Geological Society of London President's Award recipient, at the age of 29. His primary research interests are in high-pressure experimental petrology, while out of the laboratory he is a keen mountaineer and artist.

## L. Lonergan

Lidia Lonergan graduated from Trinity College, Dublin, in 1988 with a first class honours degree in geology. From 1988 to 1991 she studied the tectonics of southern Spain for her D.Phil. at Oxford University. She then joined Shell Research in Holland and worked in a team developing software for modelling the formation of basins and the generation of oil and gas. Since late 1994 she has been at Imperial College, first as a Fina Lecturer and since 1996 as a Royal Society University Research Fellow. At Imperial College she has a taken a leading role in building up the 3D seismic interpretation facility and manages the interpretation laboratory. Her main research interests lie within the fields of structural geology and tectonics and she finds it particularly fruitful to use 3D seismic datasets to address fundamental scientific problems within the earth sciences. When not finding excuses to go hiking in mountain belts the world over, she maintains a healthy interest in all things Irish, Italian, cooking and reading literature.

## N. White

Nicky White has a first class honours degree in geology from Trinity College, Dublin. From 1984 to 1988, he carried out a PhD in geophysics at Cambridge. The next two years were spent working for the British Institutes Reflection Profiling Syndicate (BIRPS). Since 1990, he has lectured at Cambridge where he is now an Assistant Director of Research. His research interests are focused on extracting quantitative information about Earth processes from sedimentary basins worldwide. Much of the data he works on has been acquired by the petroleum industry with whom he has close and fruitful links. He runs a research group of 12 PhD students.

## N. R. McDonald

Robb McDonald graduated from the University of Adelaide with first class honours in applied mathematics in 1986, and in 1990 obtained his PhD from the University of Western Australia. Robb was a Royal Society Endeavour Fellow at the University of Oxford, where he did research in geophysical fluid dynamics, and was later an ARC Postdoctoral Fellow at Monash University. Since 1994 he has been a Lecturer in Mathematics at University College London, where he continues to do research in geophysical fluid dynamics.

## M. A. Saunders

Born in Tankerton, Kent, Mark Saunders studied at Southampton University where he graduated with first class honours in Geophysical Sciences in 1978. He obtained his PhD in Space Plasma Physics from Imperial College London in 1982. He was a European Space Agency Research Fellow at UCLA, and a Royal Society 1983 University Research Fellow at Imperial College London. He joined University College London in 1993 where he is now Senior Lecturer, and Principal Climate Physicist in the Benfield Greig Hazard Research Centre. Dr Saunders leads a group specialising in the long-range prediction of industry-sensitive weather and climate. He is a frequent speaker at conferences and workshops and has published more than 80 scientific research papers and articles. His recreations include running, cricket and golf.

# Index

Page numbers in italic denote figures